BIOTRON
IS THE RULER OF BIOEMPIRE ON EARTH AND UNIVERSE

A C Pullaiah

INDIA • SINGAPORE • MALAYSIA

Notion Press

Old No. 38, New No. 6
McNichols Road, Chetpet
Chennai - 600 031

First Published by Notion Press 2019
Copyright © A C Pullaiah 2019
All Rights Reserved.

ISBN 978-1-64546-817-2

This book has been published with all efforts taken to make the material error-free after the consent of the author. However, the author and the publisher do not assume and hereby disclaim any liability to any party for any loss, damage, or disruption caused by errors or omissions, whether such errors or omissions result from negligence, accident, or any other cause.

No part of this book may be used, reproduced in any manner whatsoever without written permission from the author, except in the case of brief quotations embodied in critical articles and reviews.

TRANSLATION

Dr. C.V. NARASIMHA MURTHY
M.Sc., Ph.D., M.Ed., D.H.Ed.., P.G.D.J., D.P.R.A
P.G.E., R.A. (U.S.A)
Head of the Department of Zoology
J.B. Degree & P.G. College, KAVALI

My gratitude To

Late M. Ananda rao
Retired Lecturer, Zoology
NELLORE

INDEX

Foreword   11

Preface   13

PART – I
ASTROBIOLOGY

Atomic Age – Cell Age – Genomic Age	19
Origin of Astro-Bio-Physics and Its Necessity	20
New Concepts Derived with Coordination of Astronomy and Biology	21
New Waves Explaining Relationship between Abiotic-Biotic Systems	23
Proton, Which Produces Bio-Waves (Protomagnetic Waves)	24
Birth of a Star	26
Identification of Ultra-High-Cooled Stars and Their Composition	29
Existence of Non-Luminescent Star families	30
Identification of Double Convex Celestial Body and Its Composition	35
Process of Biogenesis	40
Object – Image Relation in between Abiotic and Biotic Systems	41

PART – II
ASTROZOOLOGY

Chapter 1   49

Role of Bio-Waves (Protomagnetic Waves) in the Life of an Animal from Birth to Death   49

Growth – Retention – Decrease of a Creature Due to Related Planet Rotation   53

Astro Mathematical Data Related
to the Bio-Mass of a Creature   56

Chapter 2

Size of an Organism of a Species Responsible to Bio Waves Frequency of Related High Cooled Celestial Body   59

Formation of Bodies of Creatures Due to Bio-Waves
Emitted by Celestial Bodies   59

Chapter 3

Sexuality in Plants, Animals, and Lower Organisms   63

Areas of High Cooled Celestial Body (Planet) that Produce
of Male, Female and Neuter Genders   63

The Formation of Non-Gender Variation in Plants By Bio-Waves   65

Chapter 4

Biofield Mode of Non-segment and Segment Structure Animals by Bio Field Type 1 and 2   68

Chapter 5

Shape and Size of an Organism   75

Effect of Massive Body (Non-Luminous Star) on a Planet   75

Animal Formation and Birth   75

Chapter 6

Method of Formation of Important Organs Like Eye, Ear, Nose, and Mouth by Fundamental Forces   83

Formation of Eyes, Nose, Mouth and Ears
by Four Fundamental Forces   83

Reasons for Not Having Parts such as Eyes, Nose
and Mouth in Plants   86

The Bio Field Structure of a Plant   88

Chapter 7

Concept of Biotron   91

Concept of Biotron   91

Formation of Body Parts such as Tail and Horns in Animals by Resultant Force   93

Chapter 8

Reproduction Regulated by Bio Wave Mechanism   96

Assimilation of Biotron by Genome   98

Self-Replication in Chromosomes by Biotron Power   101

Meiosis in Germinal Cells Due to Null Point Area of Biofield   102

Recombination of DNA in Chromosome Inconsonance   103

Conjugation Conducted by Biofields   104

Nucleolus or Plasmosome, Which Holds the Biotron?   110

Spiralisation and Condensation of Chromosomes Due to Resonance   112

Formation of Biotron Due to Credited Power by Spiralisation of Chromosomes   114

Dibraglee Doctrine – Interations of Matter and Waves – Birth of Biotron   115

Enhancement of the Importance of Nucleolus with Imbibition of Biotron   117

Doppler Effect   119

Physical Properties of Biotron Directing the Fertilization   120

Inner Force Lines of a Bio Field Causing Parturition   121

Bio-Rays Causing Meiosis and Relevant Mathematical Data   123

X-Genome Spermatozoon Avoiding Fertilizing with Y-Genome Spermatozoon   125

X-Genome Ovum Is Able to Fuse with X-Genome Sperm Cell   127

XX Is Able to Form a Female but YY is Unable to Form a Male   128

Physical Principles that Suppress X-Genome in Males	129
Females Losing Dominance in Nature Due to Gravitational Force	132
Dominant and Recessive Nature of Chromosomes Due to Change in Biotrons	132
Parturition in Marsupials – Control of Bio Field	134
Biofields Causing Variations Like Oviparity and Viviparity	135
Bio-Waves Causing Slight Variations Between Spermatogenesis and Oogenesis	135
Biotrons Causing Size Variation – Why Sperms are Smaller than Ova	139
DNA Loops in Lamp Brush Chromosomes – Role of Biotrons as Causative Force	140
Explanation on Active and Passive States Experienced by Organisms	141
Reason for Highly Intelligence of Human Beings among the Rest of Animal Species	153

Foreword

The biotron is responsible for the entire bio-empire on Earth. There are crores of nonluminous highly cooled solar systems in our galaxy. In this, some solar systems have a special facility in nature. Solar systems generally consist of planets and satellites. Each and every sun of these solar systems create a plant species, and every planet and satellite creates a herbivorous and carnivorous species on the Earth.

This creating and controlling nature is achieved by highly cooled solar systems, through bio-waves. These bio-waves contain energy pockets, or quantas. These are called biotrons. Every cell has this biotron in its nucleolus. This concept of the biotron is based on physics and astronomy.

The book gives the reader a full explanation and clarity about the existence of variations in all levels of life: kingdom, phyla, genus, species, organisms, organs, tissues, cells, nuclei, and genomes.

If you study this book, you can gain more knowledge about evolution, and you will be in a position to completely reject the evolution theory proposed by Charles Darwin.

Astrobiology is still in its infancy compared to the breadth and depth of biological research in the world. This subject receives relativity little attention from the biological as well as astronomical societies. This lack of research awareness is a major oversight, considering astronomical bodies located far away from the Earth. The knowledge about ultra-highly cooled celestial bodies and their radiated bio-waves are tremendous. The book without references means it is gifted by God. At the same time, the author is not a religious, spiritual devotee of God. He is a highly rational, motivational, and diligent person.

I am therefore delighted to see the completion of this book. This book will provide researchers, educationists, students, and scientific societies an intuitive understanding of astrozoology and astrobotany, and will create a true fascination for this life process. My appreciation goes to Sri A.C. Pullaiah,

Rtd. Headmaster. I congratulate Dr. C.V. Narasimha Moorthy, Head of Zoology, J.B. Degree and P.G. College, Kavali for translating this book from Telugu to English.

Sir Gregor Johann Mendel did not have an identity for his voluble contribution to genetics in his lifetime. After a century, our world recognized and reinvented his theories. Then, he was honoured as Father of Genetics. Likewise, the basic principles and existence of bio-waves and biotrons are not known to the world at present. This may be reinvented within a short time due to the rapid growth of science. Then, our author A.C. Pullaiah will be honored by the world's scientific community.

Dr. H. Rama Subba Reddy,
M.Sc., B.Ed., M.Phil., Ph.D.,
Asst. Professor in Zoology,
Govt. Degree College,
Banaganapalli.
Kurnool (Dt.), A.P.

Preface

It is doubtless to state that biology is one of the most developed sciences of today. With the developing knowledge of modern sciences like chemistry and cellular biology, biology is being developed rapidly. It's a known fact that the protoplasm's structure and function are recognized at the atomic level due to the enormous development of this branch of science.

Though it has been developed so quickly, forming relationships with many other branches of science, biology in all its knowledge has not been able to explain the basis for life. In other words, the knowledge and volumes related to biology enable people to know the characteristics and features of living beings, but you cannot find a single line that could define or explain what life is. Or what the basis for life is? It is like a thirsty man searching for water in mirages.

To overcome this distressed state of knowledge, the basis for life is nothing but a biotron—an "energy pocket" like a photon or a quanta, found in electromagnetic waves. This conclusion has been delivered after many observational studies and essential experiments. This volume not only explains the various bio-diversified forms that live on the Earth but also gives information on how a living being is formed of and controlled by biotrons. All the details and knowledge of this have been put or incorporated in this book, titled *The Biotron: An Explanation of Life,* but to avoid confusion from the diversified forms of biology, the content of this book has been covered under five different themes: Astrobiology, Astrozoology, Astrobotony, Astro Microbiology, and Modern Astrobiology.

The knowledge that we gain in this book has been developed in three stages. The first deals with recognizing the origin and existence of the biotron, with the help of the knowledge of both physics and astronomy. Explaining how this biotron forms and controls living beings in the classification orders of kingdoms (genus) and species (living beings, organs, tissues, cells, nuclei, and genomes) is the second level. The third deals how this biotron controls and performs the characteristics of life forms: birth;

growth; death; feeding; reproduction; heredity; differences in body sizes, food habits and shelter-sex; and so on.

As there is no standard knowledge volume that states and explains a single line about what life is, this book has made it possible to know about the basis for life and explains life at all levels. Really, to bring out this volume has undoubtedly been a Herculean task.

The information in this volume is new even in the imagination, and many abstract and technical terms may not be found in present-day science books. While explaining the concepts, only a little preliminary information is given, and the rest will be learnt on studying other chapters. Only then can one have detailed and complete knowledge of the biotron. This book presents the main basic principles of Astronomy, Physics and Biology in a simple way. It also contains simple out line diagrams to make everyone understand easily. Now a days every one is having an Android mobile with him. If readers come across any unknown terminology or concept while reading, they may take the help of Google search for instant additional information.

As the author of this volume in which I try to explain life in a different angle, I humbly request you to ignore any errors and receive the essence of reality, which in turn is an encouragement to me.

A.C. PULLAIAH
Cell No: 9948773983
Email ID: acpullaiah@gmail.com

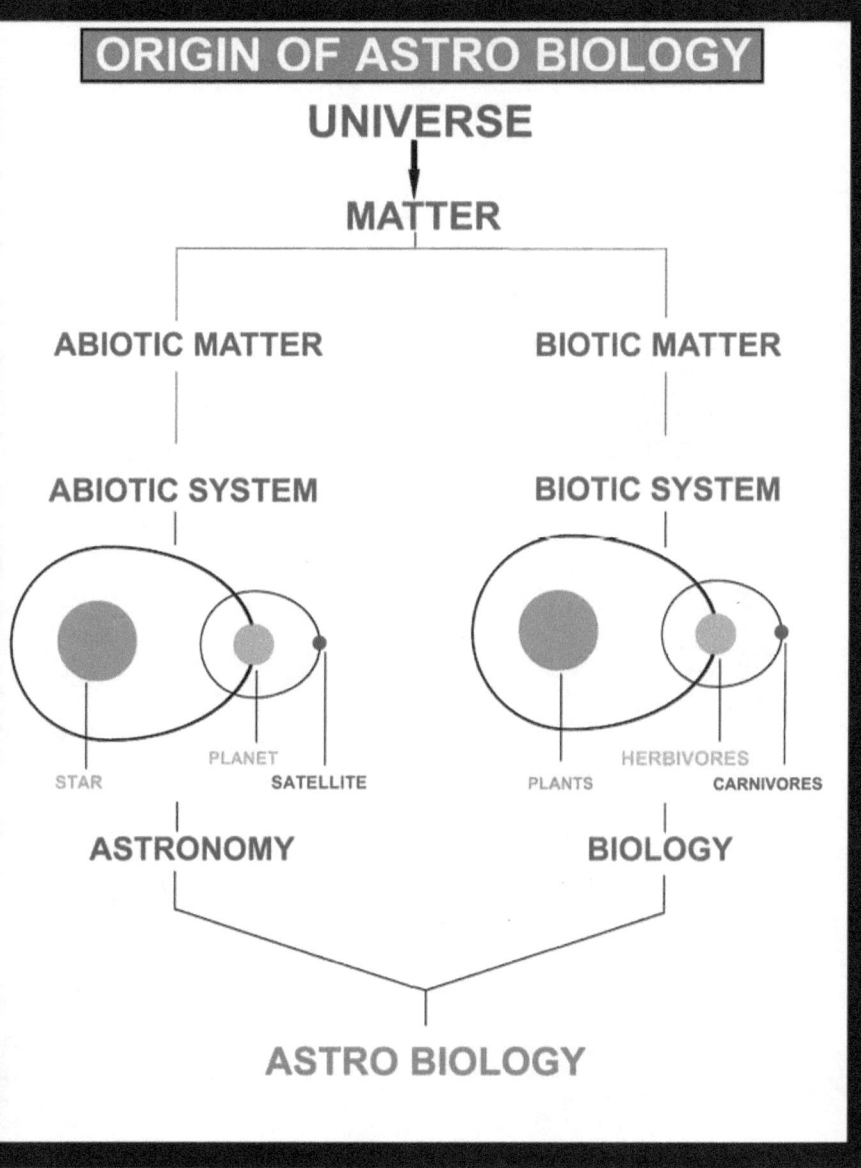

PART I

ASTROBIOLOGY

Astrobiology

ATOMIC AGE – CELL AGE – GENOMIC AGE

Much progress has been made in the field of atomic physics after the industrial revolution in Europe. The atom was studied in detail, we prepared the atom bomb, and we entered the atomic age.

In the same way, we are able to study, in detail, the cell, which is considered to be the basic unit of life and living systems. In 1953, Watson and Crick proposed the DNA model, and the basis for inheritance and genetics were found. With the advent of endo-nuclease enzymes in 1971, the knowledge of cytology was enriched. Now, we can control various biochemical events that occur in a cell.

Through the achievements in the field of biotechnology, we can now analyse chromosomes, genes, and DNA, and can produce hybrids, cybrids, mAbs, clones, and transgenics. In the human genome project, the entire human genome was analysed and decoded to various gene structures. Now, we are proud to enter the genomic age from the atomic age.

The nature of matter or materials in living beings is studied in detail. In spite of all these developments, we are unable to understand the nature of the vital force or energy that is present in organisms, which is responsible for life characteristics. It is something like a coin with two faces. One face is matter and the other face is the vital force or life. Even though we are able to understand the nature of matter, we do not have any idea how non-living material got life and its characteristics.

Thus, the complete nature of life is not properly understood. To understand the nature of life, we have two approaches: Earthly and extra-earthly views on two sides of the bio-coin.

1. **Terrestrial Approach:** According to the terrestrial approach, various physico-chemical factors that have been present on Earth before 3,000 million years ago are the basis for the genesis of life. Reduction

factors in that environment are considered to be the basis for life on Earth. In support of this theory, in 1952, Stanley Miller conducted experiments in the University of Chicago. In his experiments, he could synthesize amino acids, sugars, proteins, and nucleotides, which are all important for life.

Later, a Russian scientist called A.I. Oparin proposed his Coacervates theory, which explains the formation of carbon polymers and development of colloidal systems and the genesis of life. But none could find the nature of vital force or energy that exists in the living systems and is responsible for interactions between these molecules and the genesis and perpetuation of life.

2. **Extra-Terrestrial Approach:** The extra-terrestrial approach explains that the factors responsible for producing life might have reached the Earth from the celestial bodies through space. Huge quantities of material reach the Earth, from asteroids, planetoids, and comets, through meteors. Along with them, some fundamental living organisms might have entered the Earth and later evolved into different species at different stages under congenial conditions. In 1914, we could trace the debris of comet Thenguska in the Siberian desert, Russia. We could not find any evidence of living beings in any planets or asteroids or celestial bodies so far. We could only find traces of amino acids and carbon polymers.

ORIGIN OF ASTRO-BIO-PHYSICS AND ITS NECESSITY

The above two theories fail at explaining life and its genesis and characters. Hence, there is a need to find the factors responsible for life and its existence. This made me investigate the influence of celestial bodies on factors responsible for life, which has ultimately developed into Astrobiology.

With the advent of science, technology, and instrumentation, we can explore the universe up to 12,000 million light years. These studies show that the entire universe is filled with matter and its radiation. Matter is present in stars, planets, satellite, comets, meteors, and interstitial space of galaxies in the universe, and black holes. Thus, the entire universe is filled with matter.

In the same way, living beings on Earth are also made of matter.

Conclusion

1. Life may be the internal content of matter, or
2. Life may be the interaction of matter with certain waves or radiation.

So far, there is no evidence to prove the existence of any particle or structure present in the atom that is responsible for life. Hence, it can be deferred. According to physics, matter and energy are synonymous. So the interaction of specific radiation or waves might have led to life in non-living matter.

For example, an inert radio can produce dialogues, sounds, and speech after receiving specific radiation or waves from the space. In the same way, matter might have received specific radiation from extra celestial bodies that caused it to exhibit characteristics of life. Hence, we should make efforts to study life in this new direction. In pursuit of this, we should study the radiation or waves that are present in the universe.

NEW CONCEPTS DERIVED WITH COORDINATION OF ASTRONOMY AND BIOLOGY

Comparison of Astronomical Knowledge with Knowledge in Biology

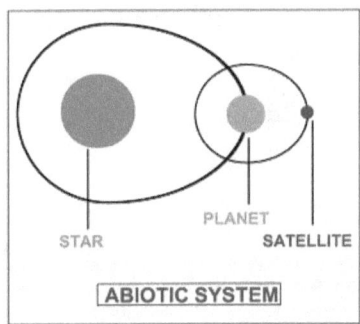
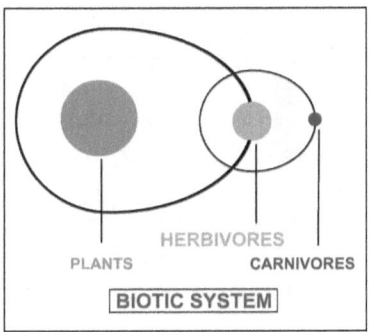

At present, all matter in the universe is of two types: Non-living matter that is in celestial bodies.

Living matter with vital force. Non-living systems contain stars, planets and satellites.

Satellites revolve around planets.

Planets revolve around stars.

Living matter is basically in three forms. They are plants, herbivores, and carnivores. Some of the comparisons and contrasts between non-living systems and living systems are given below:

S.No.	Non Living System	Living System
1.	Stars are self-luminescent.	Plants are autotrophic.
2.	Planetary motion is under the influence of stars.	The herbivores kingdom is dependent on the plant kingdom.
3.	Satellites revolves around planets.	Carnivores are dependent on herbivores.
4.	Celestial bodies (stars, planets, satellites) have their own rotational periods.	Living beings (plants, herbivores and cornivores) have their own life periods.
5.	Stars appear to be static.	Plants are sedentary.
6.	Planets revolve around stars.	Herbivores move towards plants for nutrition.
7.	Satellites move around planets.	Carnivores move towards animals for their nutrition.

We can deduce the following conclusions after studying the above two systems:

1. There is a relationship between stars and plants, planets and herbivores, and satellites and carnivores.

2. Every star has its rotational period. In the same way, every plant species has a definite living period.

3. This provokes the idea that each plant species is related to a specific star. For example, all neem plants are represented by one star, all tamarind plants are represented to one star, and so on.

4. All herbivore species have a specific longevity. In the same way, all planets have specific rotation periods.

5. This means all elephants are represented by one planet, all horses are related to a specific planet, and all human beings are related to a particular planet.

6. Every carnivorous species has a particular living period. In the same way, every satellite has a specific rotation period.

Ultimately, the above assumptions can lead to the conclusion that each star is related to a plant, each planet is related to a herbivore, and a satellite is related to a particular carnivore species.

If these assumptions are true, there should some relationship between remote stars and plants or remote planets and herbivores. Actually, all stars and planets are spread over the universe at millions of kilometres or light years.

If there is any relationship between celestial bodies and living beings on Earth, it should only be through the waves. We know that far away space probes are controlling from the Earth by the waves even though they are more than a million kilometres away from the Earth.

NEW WAVES EXPLAINING RELATIONSHIP BETWEEN ABIOTIC-BIOTIC SYSTEMS

If celestial bodies have a connection with life on Earth, then we should know what types of waves are used to control life.

Today, modern physics recognizes the different types of radiations that are present on Earth and has categorized them under the electromagnetic spectrum. All of these are in the form of waves, such as TV, infrared, ultraviolet, x-rays, gamma rays, and visible colour waves. Out of these, ultraviolet rays, x-rays, and gamma rays have higher ionization energy, which is destructive, rather than constructive. So these waves cannot be used to construct life. Radio waves, TV waves, and infrared waves have neither construct nor destroy life. But plants are able to utilize photons from solar energy and are able to synthesize carbohydrates and plant structures. So there must be some amount of constructive nature.

These light waves are available for only twelve hours on Earth, and during the other twelve hours, they are not available. Hence, the possibility of using these photons for constructing or controlling events in living beings is not possible. Apart from this, some organisms live in the bottom of the sea, where light cannot penetrate.

From this, we can conclude that the visible spectrum of light cannot control the events that occur in organisms.

Now, we should think about some unknown radiation that might be causing the link between matter and the vital energy that is present in life.

We should explore the possibility of unknown radiation from celestial bodies influencing life on Earth.

Physics Explains the Nature of Waves

1. When matter vibrates, waves are produced.
2. When particles are charged while producing waves, the resultant waves are also charged.

From the above concepts, waves are basically of two types. They are:

1. Waves with charge.
2. Waves without charge.

The well-known sound waves and weak gravitational waves are not charged. They do not have any constructive energy and they will become weak over a distance. That is why charge-less sound waves and gravitational waves are useless for constructing life.

PROTON, WHICH PRODUCES BIO-WAVES (PROTOMAGNETIC WAVES)

If you take charged waves into consideration, they are produced by charged matter or particles. Basically, all the matter in the universe is made of atoms. Each atom is made up of electrons (negatively charged), protons (positively charged), and neutrons (no charge).

When negatively charged electrons are vibrated, radio waves and TV waves are produced. Normally, within an atom, electrons will move in different energy levels. When an electron receives energy from outside, it jumps to the next energy level (excited state) and comes back to its normal level (ground state). In this process, some energy will be liberated, which will be in the form of visible light rays, ultraviolet rays, or X-rays. But we study previously these negatively charged electrons emitted waves cannot bring interactions between non-living and living systems.

Protons are positively charged and are bound in the nucleus with a high amount of binding energy. So vibrating a proton within a nucleus is very difficult. We can isolate protons and make them vibrate and can generate waves. When hydrogen is in the plasma state, free protons are available. Such protons can be vibrated in a positively charged field, and waves can be produced.

A proton has 1,836.12 times higher weight than an electron and has a positive charge. So we can't conclude that waves that can be produced from protons are not like electromagnetic radiation.

The base for electrons is fermions, which are indivisible and stable. But protons are made up of quarks, which are divisible and stable.

If they produce waves, they should be different from electromagnetic radiation and can be called positive magnetic waves, or protomagnetic waves.

In an atom, electrons will revolve around the nucleus in quantified orbitals of definite values. When they receive energy from external sources, they jump to higher orbitals and return to the ground state. The difference between these two energy levels will be in the form of electromagnetic radiation. If the electrons don't receive any energy from external sources, they will be revolving only in that energy level. If we see that an electron doesn't receive any energy from outside, and when we remove the energy by decreasing the temperature to absolute zero (–273.16 °C), the electrons will move in pairs. These are called Cooper pairs.

When matter is brought near absolute zero, or to very low temperatures, then the electrons lose energy and move towards the nucleus. Due to this, the angular velocity of electrons increases, which ultimately leads to angular momentum. At this state, there will be pressure on nucleus and its protons. To neutralize the effect of angular momentum, the protons start vibrating. Thus, protomagnetic waves are produced.

We noticed the hypothesis of producing protomagnetic waves, but we have to prove it experimentally and identify them physically, and also have to record the properties and others of them. But it is a matter of money, big laboratories, and support from government or private institutions. If this is done, it will start a new innovation in physics.

Here, importance is given to proving how these protomagnetic waves are responsible for life and creating and sustaining the millions of living beings in Earth's Biosphere.

In my view, there should be a separate book with the experimental information to prove the protomagnetic waves' existence and record their properties, as all of these are not given in the Astrobiology section.

All these hypotheses are to find the type of waves that cause links between non-living matter and life.

The protomagnetic waves which resonated protons will be produced at near absolute zero—at −273.16 °C. In order to produce such waves, there should be some material nearer to absolute zero.

Actually, the average temperature of the universe is 2.73 K, or −270.42 °C. This is a very cold state. Much matter is available in the universe at this temperature. In our solar system, Zovenian planets—Jupiter (−123 °C) Saturn (−180 °C), Uranus (−218 °C), Neptune (−228 °C) and Pluto (−230 °C)—are present in severely cold conditions.

Scientists are of the opinion that comets are formed from the Oort clouds far behind the sun in the solar system. These comets are at a low temperature (less than −230 °C) because they are far away from the Sun. The distance between a star and another star will be measured in light years.

From the above information, we can conclude that there is matter with very low temperatures. That matter might be in stars, planets, or satellites. Under any circumstances, stars cannot have low temperatures, because the outer temperature of the star will be thousands of degrees and the inner temperature at the core will be around millions of degrees. For example, the surface temperature of the Sun is 6,000 °C, and at the core it will be around 3,60,00,000 °C. If a star is glowing, it must be at a minimum of 20,00,000 °C. Only then will thermonuclear reactions occur and energy will be liberated. So there is no possibility of having near-absolute temperature in any star.

Previously, we proposed the concept that one star relates to one plant species, one planet relates to one herbivore, and one satellite relates to one carnivore. This will go wrong if all stars have higher temperatures. To circumvent this problem, we should explore any other possibilities from astronomical sciences.

BIRTH OF A STAR

Genesis of Stars

According to Astronomy, there are 150 billion stars in the Milky Way, which make it a galaxy. These stars are similar to our sun. According to an estimate, 200 billion such galaxies exist in the universe. It is learnt that stars originate from the hydrogen molecular state clouds (gravitational collapsed nebula fragments) present in the galaxies. Due to differential rotation of galaxy, these gaseous clouds, the molecules in the cloud come nearer and the volume increases. After attaining a certain amount of mass, it starts attracting the surrounding matter, and the volume of the star increases. As the size increases, the mass will be increased. At this juncture, the gaseous molecules will be exposed to friction and severe heat will be generated. At 20 million degrees, thermonuclear reactions will occur in hydrogen atoms, converting them in to helium and liberating a huge amount of energy. From then, the gaseous cloud glows like a star.

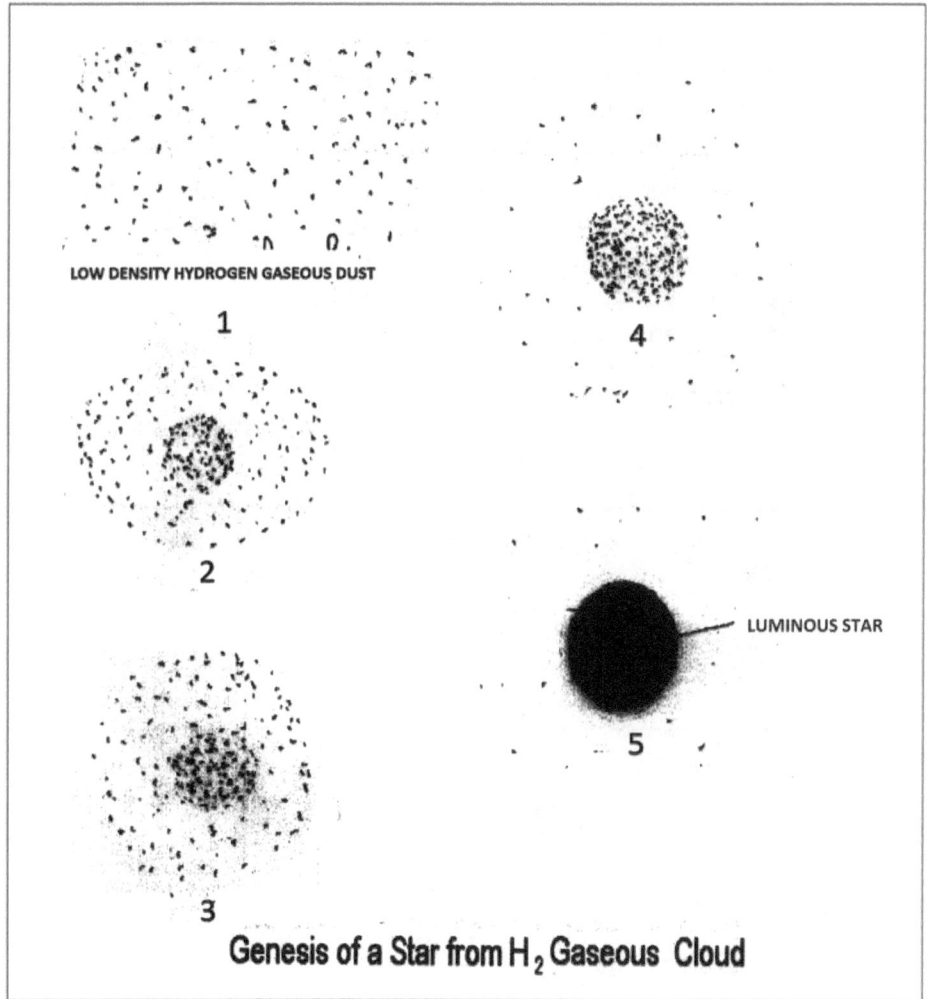

Genesis of a Star from H_2 Gaseous Cloud

It is difficult to tell whether all stars have planets around them but Astronomy knowledge clarifies this as follows: Due to the differential rotation of galaxy and by the surrounding stars influence, within the gaseous cloud (that is present within the galaxy), gravitational force is generated, and the cloud attains the shape of a wheel and rotates around its axis. In this state, small whirlpools (gravitational centres) will be produced at some places within the wheel. At the centre of the wheel, more mass will be accumulated, and ultimately it becomes a big sphere. Due to friction between molecules in this sphere, heat will be generated. When this heat reaches 20 million degrees, thermonuclear reactions start, and the sphere starts glowing like a star.

In the gaseous wheel of the cloud, there are some gravitational centres where the matter is accumulated in lower quantities than in the centre. As the matter in these spheres is lower, there will be less friction, and the temperature produced doesn't exceed 20 million degrees. Thus, there will be no thermonuclear reactions in the sphere, and it will not glow. Thus, planets form around the stars. At the centre of this cloud, a large amount of mass will be present, which makes the planets move around the star. If some substance is present around these planets, it will be called a satellite and will move around the nearest planet. Thus, similar types of stars and planetary systems like solar system will be present in the universe.

There is a reason small planets like Mars, Venus, Earth, and Mercury are formed nearer to Sun and large planets like Jupiter, Saturn, and Uranus in middle and at planets like Neptune, Pluto at the end. Moreover, specific distances between different planets can be explained by Astrophysics knowledge.

If we consider the genesis of stars like solar systems according to this theory, it will give a scope for new thought, as shown below.

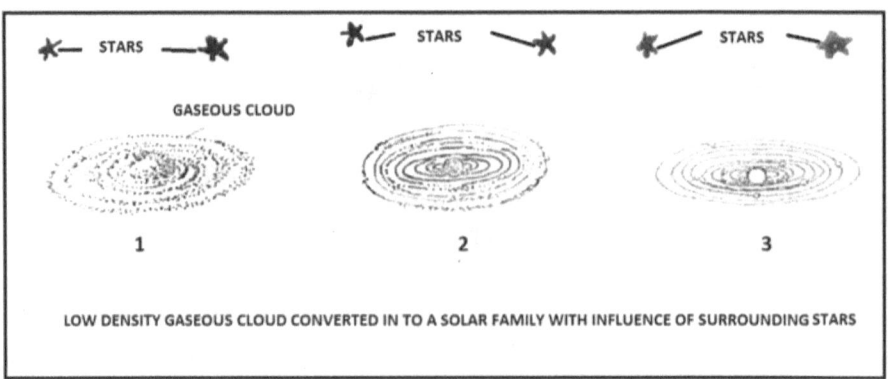

IDENTIFICATION OF ULTRA-HIGH-COOLED STARS AND THEIR COMPOSITION

Stars originate from the hydrogen clouds present in the galaxies. Due to differential rotation of these gaseous clouds, the molecules that are in the cloud come nearer to each other and the volume increases. After attaining a certain amount of mass, it starts attracting surrounding mass and volume of the star increases. As the size increases the mass will be increased at this juncture the gaseous molecules will be exposed to friction and heavy heat will be generated. At 20 million degrees, thermonuclear reactions will occur in hydrogen atoms and convert them to helium, and huge amounts of energy will be liberated. From then, the gaseous cloud glows like a star. In a gaseous cloud, if less mass accumulates at the centre of the wheel, it becomes spherical. Due to friction between molecules in this sphere, heat will be generated. If this heat doesn't reach 20 million degrees, there will be no thermonuclear reaction. So the sphere cannot glow like a star. At different gravitational centres, the masses around this star will become planets. As there is no thermonuclear reaction in this star, they will be cooled slowly. Thus, the temperature decreases in these non-luminescent stars. As there is no heat production, the surrounding planetary system will also be cooled.

Eventually, all non-luminescent stars and planetary systems will be cooled and reach absolute zero. At absolute zero, due to the increase of angular velocity of electrons, protomagnetic waves will be produced. Thus, the entire non-luminescent star with planetary system will start producing protomagnetic waves.

So far, we have explored the possibility of the existence of non-luminescent stars and planetary systems according to one of the star genesis processes, which supports the existence of living and non-living systems and the presence of plants and animals. It gives support for the existence of non-luminescent stars and the production of protomagnetic waves.

Finding the genesis of non-luminescent stars is one aspect, and we have to trace the presence of these non-luminescent stars and planets in the universe, which is spreading over hundreds of crores of light years. We have to find out their number and distance in our universe.

EXISTENCE OF NON-LUMINESCENT STAR FAMILIES

According to the Big Bang theory, a heavy explosion caused atomic matter that is in the shape of an oval to be scattered thousands of light years into space. The bits of matter that were spread into space due to self-rotational force started condensing and gradually formed different islands. Each island is eventually developed into a galaxy. Within a galaxy, millions of stars and planets are present.

If the mass of a star or planet is less than 100 times the mass of the sun, it cannot produce 20 million degrees. To conduct thermonuclear reactions, there must be a minimum amount of mass. In the case of Jupiter, its mass is less than the 1/100 of Sun's mass. That is why there is no thermonuclear reaction in it. So it is not glowing.

According to the Big Bang theory, the bits of matter present in the galaxy started rotating due to self-gravitational force. Due to a gradual increase in velocity, the matter present in the island started condensing and increasing its volume. Due to the increase in size, the gravitational power increased and started attracting matter from its surroundings, ultimately emerging as large spherical body. When the size increased, thermonuclear reactions started, leading to the production of luminescence like stars. Our solar system also developed like this in the Milky Way galaxy. In this Milky Way galaxy, more than 150,000,000,000 luminescent stars are present. In the same way, more than 200,000,000,000 galaxies are there in the universe. In addition to that, our Milky Way galaxy is moving along with other 23 galaxies in a cluster of galaxies or a local group of galaxies.

There is no vacuum between galaxies. This is filled with less densely distributed gas in the form of hydrogen molecules and dust particles. At the periphery of the galaxy is also having the thin layer of gaseous dust.

In the beginning, our galaxy was thought to be in an elliptical shape. Since this galaxy is in the constellation of other galaxies, it is influenced by other galaxies. Due to its increase in rotational speed, it changed its shape from oval to spiral. Because of this, the stars present in the galaxy migrated into the arms of the spiral. The low density of gaseous dust remains in between the spiral arms of the galaxy. When these gaseous clouds are influenced by the surrounding stars in the spiral arm, the low density of this gaseous dust starts moving like wheels (clearly explained in next paras). As in the case of star genesis at the centre of the moving gaseous dust cloud, gravitational centres are formed and start accumulating matter at the centre. Here, more matter will be accumulated at the core instead of the periphery. This matter should take the shape of a sphere.

The low density of gaseous dust at the core cannot produce heat due to friction due to a lack of mass (less than 1/100 sun's mass) there will be no thermonuclear reaction and thus they are nonluminous stars. Along with non-luminescent planets, satellites are also formed at different gravitational centres around the non-luminescent stars. Thus, the entire system looks like a non-luminescent solar system. Within a galaxy, the distance between one arm and another is more than hundreds of light years. There are crores of stars in an arm of a galaxy. In the same way, there are plenty of low density gaseous clouds that will be converted into nonluminous solar families with planets that are non-luminescent. Thus, there is the possibility of having nonluminous star families in between arms of galaxy. In this book these nonluminous stars are also called often as ultra high cooled solar families or massive body families or ultra high cooled astral bodies or celestial bodies. The protomagnetic waves are also often called as bio-waves for easy understanding.

OUR GALAXY PICTURE

In the picture, the Milky Way galaxy is shown. In the Milky Way, most luminescent stars are present at the centre and spiral arms. They cannot be seen inside the spiral arm. There should either be a vacuum or some low-density gaseous dust. In reality, there is no vacuum in the universe. The entire universe is filled with universal dust, gases, or a thin layer of gaseous dust. The in between these spiral arms some low density gaseous dust clouds must be present. These gaseous clouds will ultimately become the non-luminescent solar families by the influence of surrounding stars and will produce photomagnetic waves, as they have low temperatures. From the above hypothesis, we can come to the following conclusions:

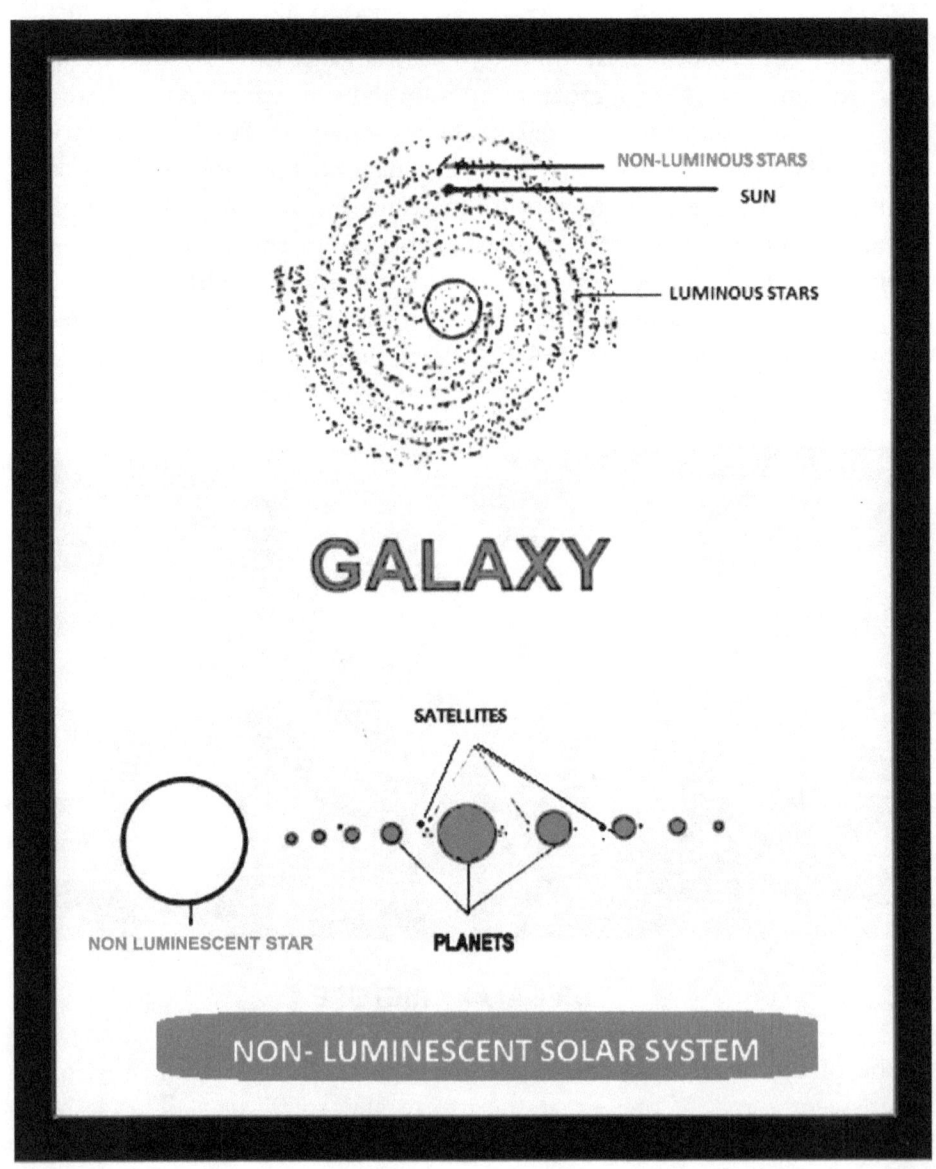

1. At the periphery of the galaxy, thin gaseous clouds are present. As they are far away from stars, they are very cool and produce proto magnetic waves.

2. Due to the change in the shape of the galaxy from oval to spiral, the low density gaseous clouds that are there in between the spiral arms of the galaxy will form non-luminescent super-cooled star families.

3. As these nonluminous celestial bodies which are far away from the stars cannot receive any energy from outside and did not able to produce energy itself. Ultimately they will be ultra-high cooled and produce protomagnetic waves.

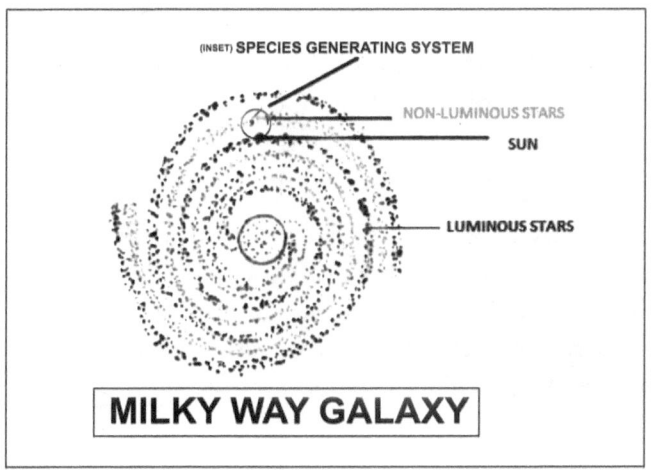

The above information will give us scope for the presence of ultra-cooled star families and the production of protomagnetic waves that are essential to establishing the relation between non-living matter and life on Earth by theoretical consideration. Now, we have to explore the possibility of identifying physical forms of images by direct or indirect methods.

Explanation

The sun has a planetary system. According the star genesis process, there is the possibility of having a planetary system on other stars. As the stars are glowing, we can see them with a telescope and take photographs to identify them. That is not possible in the case of planets and satellites, as they are not luminescent. With the advent of modern and indirect methods, our scientists identified more than 100 crore planets like Earth in the Milky Way galaxy.

As the concept of protomagnetic waves is theoretical and there are no instruments to identify and characterize such radiation coming from the universe, we cannot even see them or take photographs.

In order to establish a relation between non-living matter and life on Earth according to the current information, the existing 150 billion stars in our galaxy are useless.

Hence, we should explore the possibility of the influence of super-cooled non-luminescent star families that are present in between the arms of the spiral galaxy.

We should know the state of protomagnetic waves that are reaching the Earth from the enormous super cooled celestial bodies in the universe.

These super-cooled non-luminescent star families present in between the arms of the spiral galaxy will produce protomagnetic radiations in all directions of the universe. Only some amount of radiation will reach the Earth. The rest of the radiation will be dispersed into space.

So far, from current knowledge, we know that life exists only on Earth out of the nine planets and 85 above satellites in our solar system. Earth has limited space. But the non-luminescent star families that emit radiation or waves are infinite. Whether these waves coming from super-cooled celestial bodies are scattered, spreading, or concentrating on a particular planet, we don't know.

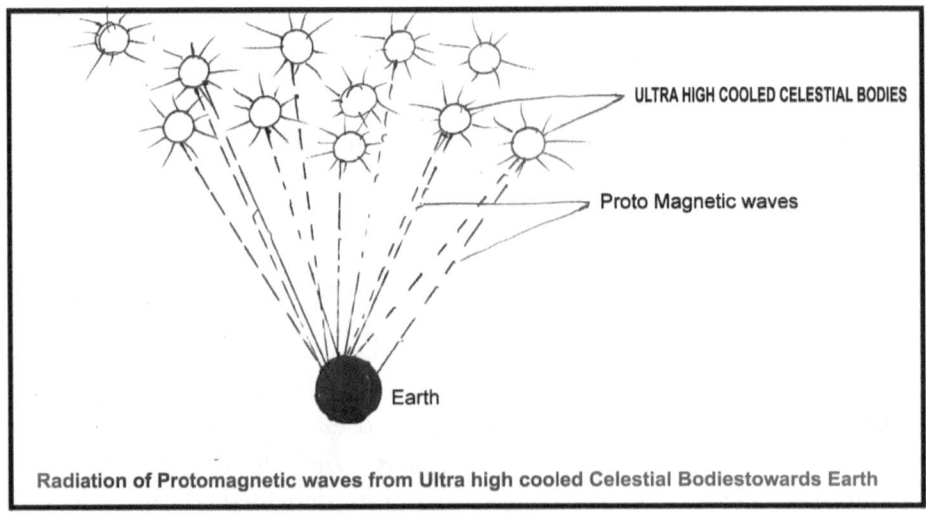

Radiation of Protomagnetic waves from Ultra high cooled Celestial Bodiestowards Earth

Though the Himalayas are wide and huge, we can take a photograph of them with a small camera. This is possible because of the lens that is present in the camera, in the same way that our eye can see bigger objects because of the lens that is present in eye. Due to that lens, large objects can be focused on the small retina of our eye.

If the entire protomagnetic radiation that is produced from the super-cooled non-luminescent star families has to reach the Earth, there should be a lens-like mechanism that concentrates these waves exclusively on Earth. So a lens-like arrangement must be present between super-cooled celestial bodies and Earth, which is responsible for life on Earth.

The following questions arise in our mind from the above hypothesis:

1. Is there any possibility of forming naturally occurring convex lenses?
2. Even if it is formed, is it transparent to photomagnetic waves?
3. Even if such a convex lens is formed, is the Earth there at the focal point of this lens?
4. The super-cooled celestial bodies or Earth the convex lens like celestial bodies are not static and sedentary. They are in motion. If so how all the waves are concentrating at particular focal point (Earth)?

In order to know the answers for the above questions, we should look again at the knowledge of astronomy.

IDENTIFICATION OF DOUBLE CONVEX CELESTIAL BODY AND ITS COMPOSITION

According to the current knowledge of Astronomy, the planets or satellites are not completely spherical. For example, in the case of Earth, the diameter of the Earth is 54 km more at the equator than at the poles. In the same way, the diameter of planets and satellites is varied. The diameter is more at meridian lines than the poles for planets like Jupiter and Saturn.

To find out the feasibility of forming convex disk-shaped celestial bodies, we should further study the genesis of stars.

Previously we discussed a theory of star genesis process. According to the gravitation theory, every meterial body or celestial body must influenced by the nearest celestial bodies due to its gravitation, so star genesis process comes under this process too. Normally, any gaseous cloud that comes under the influence of a nearest star, and starts rotating itself and attains a disk shape. For some reason, whichever star is influencing the gaseous disk, if it gradually moves away from the gaseous disk, the disk will lose the star's control, and it will remain in the disk shape instead of becoming a solar system.

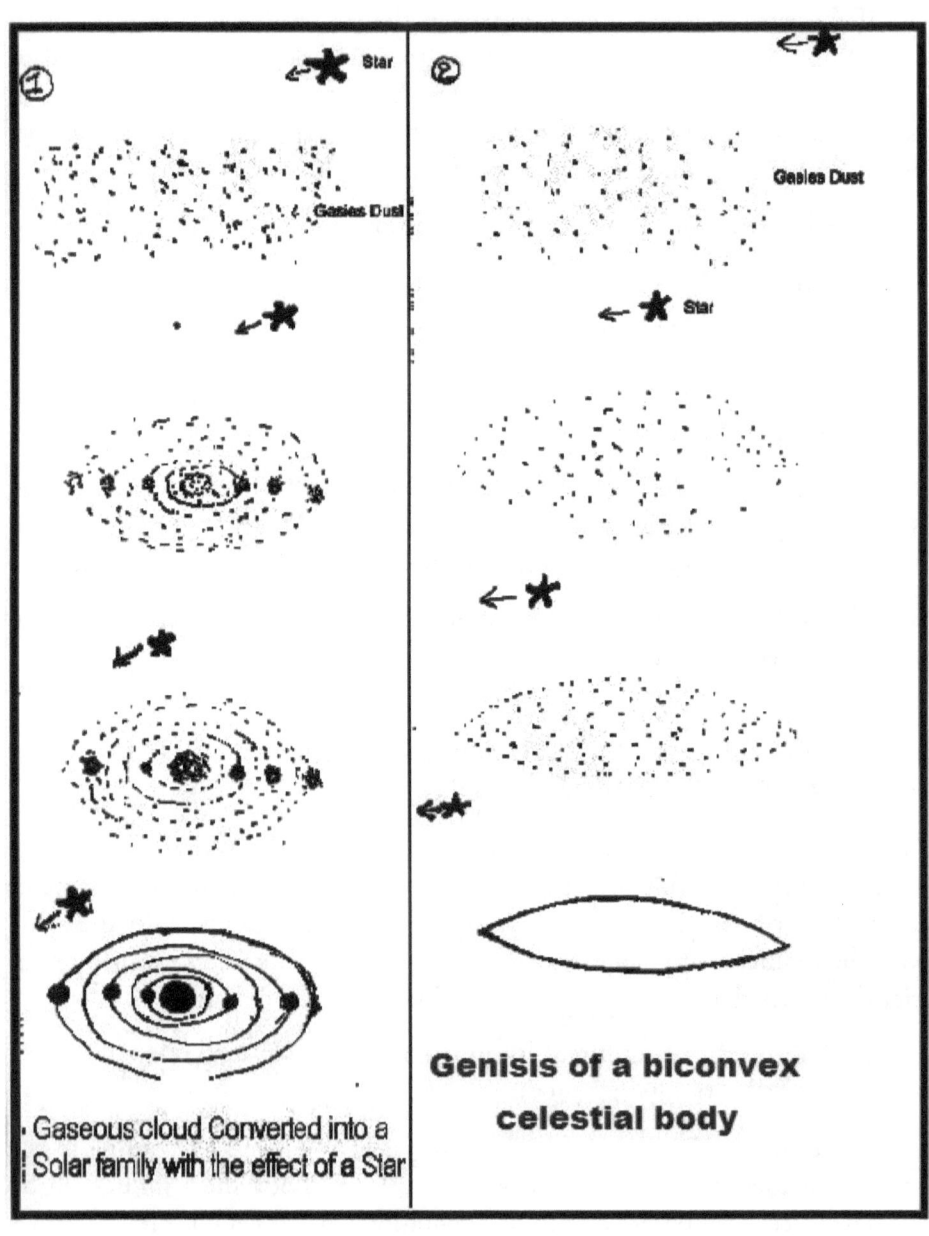

Gaseous cloud Converted into a Solar family with the effect of a Star

Genisis of a biconvex celestial body

From the above explanation, we can conclude that there is a possibility of presence of disk like bodies in the universe.

In this universe, plenty of celestial bodies are present that are biconvex in shape. A photograph of one such galaxy is given below.

In the galaxy, thousands of crores of stars are present. This giant galaxy itself is disk-shaped. So the passivity of discoidal celestial bodies is unquestionable.

(NGC 4594) SAMBREROW GALAXY

Now, we should know whether such discoidal celestial bodies are transparent to protomagnetic waves or not.

If they are not transparent, they should absorb some of such waves and radiate the rest. If these protomagnetic waves are reflected by discoidal celestial bodies, there will be no possibility for life on Earth.

So life is possible on Earth when only the disk is transparent. The transparency of a disk can be only be judged with the credibility of all astrobiology.

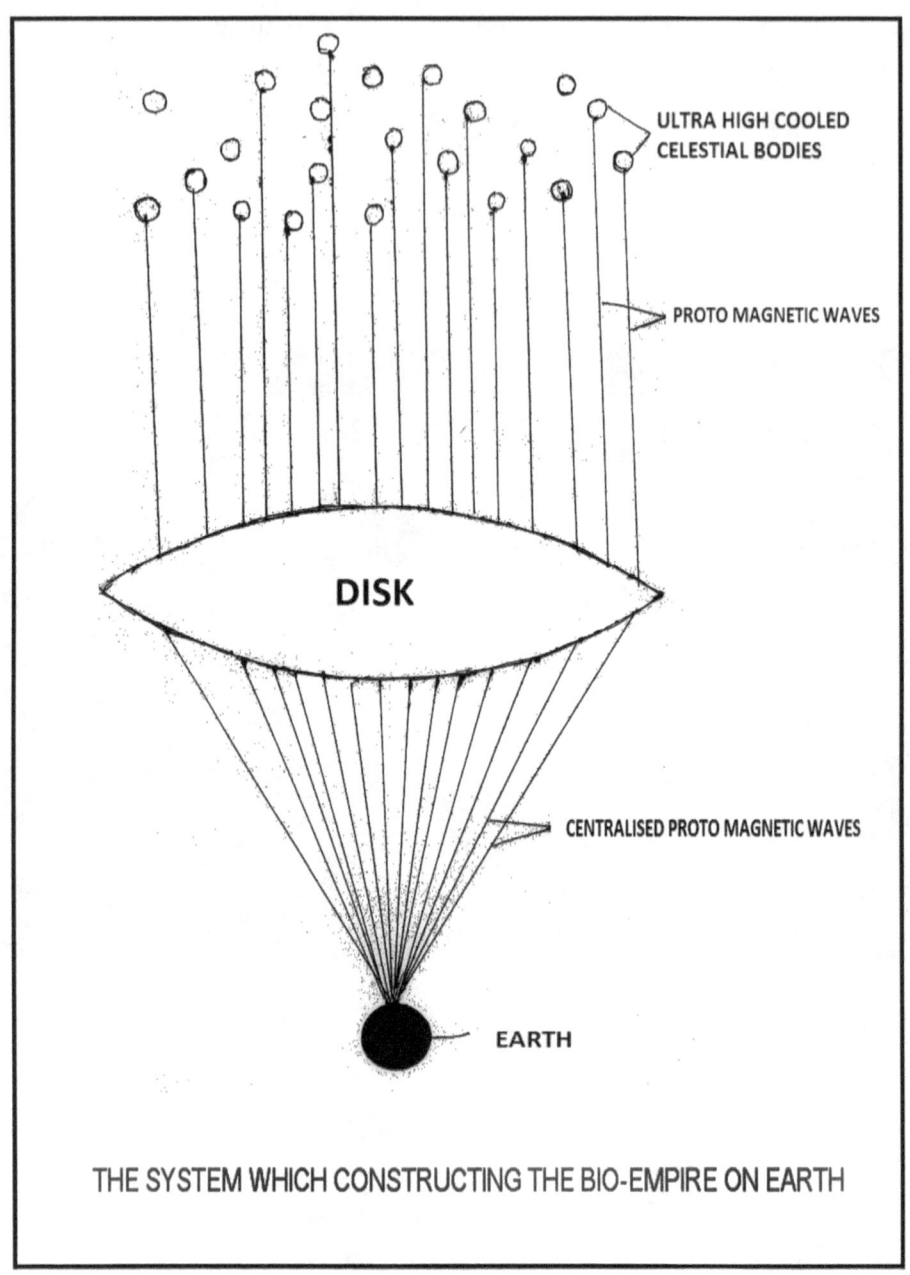

THE SYSTEM WHICH CONSTRUCTING THE BIO-EMPIRE ON EARTH

Even if you accept the existence of the disk, we have to explore the possibility of Earth being at the focal point of the disk. In addition, the disk and Earth are not static, as they are revolving around the galaxy with thousands of kilometres per hour. To concentrate the entire protomagnetic radiation onto the Earth, both it and the disk should either be motionless or have the same speed.

The following explanation will solve the above problem. When we observe the sky in the night, we feel like all stars are at fixed positions. But in reality, every star is moving at the speed of lakhs of kilometres per hour. Yet they appear to be static. The main reason for this is the actual distance between these stars and our Earth. It means that if an object is at a far-off distance, even when it is moving with high speed, it will appears to be at a fixed location. In the same way, the super-cool celestial bodies that are present in the galaxies are far from us. Because of such long distances, they also appear to be static.

The presence of the Earth at the focal point of the disk may be incidental.

If the above concepts go wrong, we can't find a relationship between non-living matter and life on Earth. The connection will break down between celestial controls of life on Earth.

There are so many factors for the existence of life on Earth. If something happens to the sun that makes it lose control over its thermonuclear reactions, the Earth will become hot and all life on Earth will die. If we were not at our present distance from the sun, and instead were as far as Jupiter, the Earth would have been at -140 °C and there would be no possibility of life. Instead, if we were in Mercury's position, the Earth would be at a temperature of 210 °C, and there would be no possibility of life on Earth. If the mass of the Earth decreased to the moon's size, all gases on the Earth might have escaped from the Earth's atmosphere, leaving no possibility of life on Earth. If the mass of the Earth increased to Jupiter's size, the atmospheric pressure might have increased to 1,00,000 times, and all life on Earth might become powder. For that matter, if there was no oxygen or carbon dioxide on Earth, or an ozone layer around the Earth, then there would be no possibility for life on Earth.

Even though there is no possibility of getting proof for the presence of a disk-like structure between super-cooled celestial bodies and the Earth, to find out the existence of a link between non-living matter and life, we have to accept the above concept as it explains life at molecular and quantum levels.

Following are the most important new concepts we have learned so far:

1. Under favourable conditions, some atoms interact with protomagnetic waves and form the life traits in matter.
2. There is a systematic relationship between non-living matter and living systems.
3. Super-cooled stars, planets, and satellites control plants, herbivores, and carnivores respectively.
4. The protomagnetic waves that are produced from the super-cooled non-luminescent star families are responsible for genesis and existence of life on Earth.
5. There is a possibility of having a disk like celestial body in between Earth and super-cooled star families.
6. The photomagnetic waves that are produced from the super-cooled star families are able to concentrate at the focal point of the disk of the Earth, which stays in the focal point and becomes responsible for production of life and its survival on the Earth.

The above concepts are framed on the discussions of physics and astronomy. But we have to find out how these waves generate life on Earth.

PROCESS OF BIOGENESIS

Genesis of Life on Earth at Gross Level

The protomagnetic waves that are generated from the super-cooled non-luminescent star families reach the Earth through the disk and influence the matter on the Earth and interact with molecules like hydrogen, oxygen, carbon, phosphorus, nitrogen, etc. This forms a field and makes the atoms a giant molecular ball which creates life in these molecules.

Due to changes in the direction, phase, and intensity of photomagnetic waves that are generated from different super-cooled celestial bodies (as they rotating themselves), their concentration on Earth will cause changes in the patterns of life on Earth. That means various developmental stages of life are influenced by these wave patterns.

This is something similar to change in an object will make change in the image of an object. In the same way, the rotation, direction, and phase of super-cooled celestial bodies will make changes in life and its diversity.

The genesis of life on the Earth by the above hypothesis is only imaginative. The existence of protomagnetic waves is imaginary. The presence of super-cooled non-luminescent star families is imaginary. So we cannot find proof or experimental evidence for them. So finding out their existence and nature by doing experiments is not possible.

Explanation

Even though there is no proof for the presence of any super-cooled celestial body or its waves and disk, the above concepts regarding them are logical and scientific in thinking. If there is a connection between two different systems, its relationship is discussed logically first and explained scientifically with mathematical derivations.

OBJECT – IMAGE RELATION IN BETWEEN ABIOTIC AND BIOTIC SYSTEMS

At present, we have identified two systems like celestial bodies and life and have discussed the relationship between them. From these two, we know something about life through biology. We don't have any first-hand information on the second one, regarding the super-cooled celestial bodies. But we came to learn about them from the genesis of Stars.

If high cooled celestial bodies are responsible for living systems, any change in the celestial system should influence the living system.

This means for any variations in the living systems, changes in the celestial bodies are responsible.

If super-cooled non-luminescent stars are objects, the life that is present on the Earth will become the image.

The image (life on Earth) that is formed can be studied through biology. The nature of the object (super-cooled non-luminescent star families) can be studied to some extent through Astronomy. Normally, objects and images are similar to total correlative systems. It means changes in the objects are reflected in the image. If you study any gradual change in the object, the same changes can be found in the images later. If you know the object and image, you can prove the identical relationship between them very easily.

Now, we have to prove the identical relationship between object and image. All variations and diversities in living systems should be explained through the changes in the celestial bodies. Only then will Astrobiology be faithful.

Millions of variations and fundamental concepts are present in living systems. All of these cannot be explained at a time. So in this chapter, we will explain some of them, and the rest will be in succeeding chapters.

Primary Factors That Find a Relationship Between Living Systems and Non-Living Celestial Bodies

Any organism that lives on Earth has life because of interaction of bio-waves with material molecules, which are emitted by ultra-cooled celestial bodies. So we have to coordinate or correlate the entire bio-empire on Earth with all ultra-cooled celestial bodies existing in the universe.

A. Celestial Bodies that Emit Protomagnetic Waves in the Universe Are:

1. Non-luminescent star families that emit protomagnetic waves THROUGH the disk in our galaxy (star, planet, satellite tri-system) (these are limited which are above the Disk).

2. Non-luminescent star families that are emitting protomagnetic waves NOT through the disk in our galaxy (star, planet, satellite tri-system). (these are unlimited in our galaxy in between the spiral arms).

3. Non-luminescent star families that are emitting protomagnetic waves in entire universe EXCEPT our Galaxy (these are also a tri-system).

4. Interstellar and intergalactic matter which is in molecular and tiny material form is hydrogen, as it is in the ultra-cooled state, emitting protomagnetic waves (single system).

B. Living Beings of the Bio-Empire on Earth Are:

1. Plants, herbivores, and carnivores in a tri-system—the first form of living beings.

2. Fungi, bacteria and virus in a tri-system, making up the secondary form of living beings.

3. Viroids, Virusoids, and Prions in a tri-system, becoming the third form of living beings.

4. In the form of water, which is the main basic for exhibiting vital force in living being, in Single System.

Correlation of A with B

1. Non-luminescent star families that are emitting protomagnetic waves are THROUGH the disk in our galaxy are responsible for the existence of the first form of main living beings—plants, herbivores and carnivores species (in tri-systems)(as we studied previously the non-luminescent star families are may be crores in our galaxy. but which are above the disk their bio waves only centralising on earth. so plants, herbivores and carnivores species are appearing limitedly on earth).

2. Non-luminescent star families that are emitting protomagnetic waves 'not' through the 'Disk' in our Galaxy are responsible for existence of the secondary form of living beings Fungus, Bacteria, Virus species (in Tri-System).

3. Non-luminescent star families that are emitting protomagnetic waves in the entire universe except for our galaxy are responsible for existence of the third form of living being viroids, virusoids and prions species (in Tri-system).

4. As interstellar and intergalactic matter, which is in molecular and tiny material form, is hydrogen, as it is already in an ultra-cooled state due to far-away stars. These are emitting protomagnetic waves is responsible for the existence of water, which is the root cause for exhibiting vital force in all living beings.

According to the above knowledge life on earth, in the form of full-fledged living being species, is limited because only limited non-luminescent star families (which are above Disk) emitted biowaves coming to earth through the Disk hence both plant and animal kingdom have only limited bio species and there is no scope for other forms of living beings of species on earth. Biology knowledge proves that this is correct.

We know that in the periodic table, from helium to uranium, all elements originate from the first element: hydrogen. We can see the neutron particles in all elements, but not in hydrogen (except deuterium and triterium), even though it is basic for all elements.

Like this, all of the bio-empire in the heavy molecular form, cell form, multicellular form, or visible organism form, exhibits vital force due to the presence of water. We know that without water, no living being can show life, because water is the primary element for life. Despite this, water itself can't show life as we notice it in nature.

Correlation in Between Abiotic Matter and Biotic Matter

According to the pyramid of mass, the number and mass of producers (plants) is highest. Next to that is herbivore mass, and carnivore mass is the least. In the same way, in non-living solar system, the total weight of the nine planets is less than the weight of the sun. In the same way, the weight of all above 85 satellites in the solar system is less than the weight of the nine planets. Out of the total mass of the solar system, 99.86% of the mass is present in the sun. The remaining 0.14% of matter is distributed among the 9 planets and above 85 satellites. In the same way, the weight of the total planets around the stars is less than 0.14%. The total planet and satellite weight of a non-luminescent star may be less than 0.14%.

From super-cooled non-luminescent stars, plants are formed. From super-cooled non-luminescent planets, herbivores are formed. From super-cooled non-luminescent satellites, carnivores are formed.

According to this, the total mass of plants should be 99.86%. All herbivores and carnivores make up less than 0.14%.

Modern biology classifies all of life on Earth into five kingdoms: Protista, Monera, Fungi, Plante, and Animalia. Among them, plants and animals are diversified, multicellular organisms. In spite of that, in all living material, birth, growth, death, cell division, and nutrition are similar.

By studying one representative species, we can deduce the general characters of similar species. In living systems, variations appear more in the plant and animal kingdoms. So a study of these two systems should be done independently. Instead of studying them independently, if we study them together, there is possibility for misinterpretation.

First, we will study variations in the animal kingdom (because the animal kingdom has crores of variations, more than the plant kingdom) through the celestial system, and after that we will study the plant kingdom. It will be easier to study it this way.

Animal Kingdom

In the animal kingdom, millions of species are present. There are certain general features among them. Therefore, by taking this into consideration when studying a representative species, we can deduce and understand their general characters. Man is also a part of the animal kingdom. We know the general characteristics of man. By taking man as an example, we will study the impact of these protomagnetic waves that are produced by the super-cool nonluminous star families. For the sake of convenience, at some places, I say bio-waves instead of protomagnetic waves.

PART II

ASTROZOOLOGY

CHAPTER 1
Role of Bio-Waves (Protomagnetic Waves) in the Life of an Animal from Birth to Death

MECHANISM OF BIRTH, GROWTH, AND DEATH OF AN ORGANISM BY BIO-WAVES

The atoms of hydrogen, oxygen, phosphorous, etc. that are present on Earth will respond to the protomagnetic waves received from nonluminous solar family bodies and form life. There will be changes in the pattern of life, such as birth, growth, senescence, and death, according to the wave power, velocity, intensity, phase, and direction of waves.

Impact of protomagnetic waves on the different stages of life such as germplasm (sperm and ovum), embryo, infancy, childhood, adolescence, youth, adulthood, old age, and death of different animals on the Earth.

1. One particular ultra-cooled celestial body is responsible for controlling one particular species of life on the Earth.

2. For example, the human species on the Earth is influenced by one particular ultra-cooled celestial body. Similarly, another ultra-cooled astral body is responsible for controlling another species.

3. Due to the rotational change of an ultra-cooled celestial body, the pattern, type, and intensity of waves produced are changed.

4. These changes are responsible for controlling various stages of life, such as germplasm (sperm and ovum), embryos, infancy, childhood, adolescence, youth, adulthood, old age, and death.

5. By studying the rotation and wave patterns of one ultra-cooled celestial body and how it influences the life of an organism and its activities, we can extrapolate the data to other species of organisms and other similar ultra-cooled celestial bodies.

6. Hence, let us study one particular ultra-cooled celestial body and its influence on various activities of its living being.

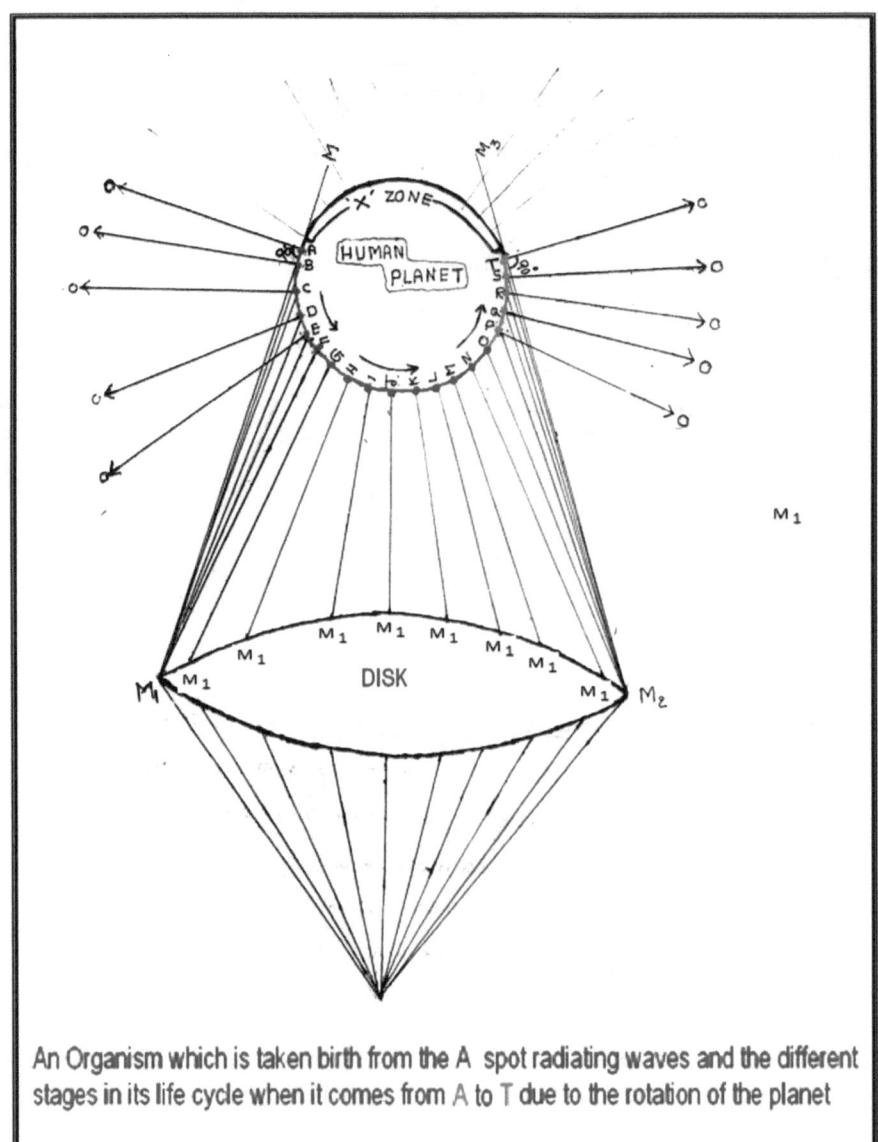

An Organism which is taken birth from the A spot radiating waves and the different stages in its life cycle when it comes from A to T due to the rotation of the planet

Let us examine one particular ultra-cooled celestial body in the above diagram. The rays that are emitted from a particular point A of the celestial body are concentrated on a particular place on Earth through a disk-like structure. The elements of that particular place in earth that are oxygen, hydrogen, carbon dioxide, nitrogen etc., absorb the energy of radiated bio waves from point A and form a single cell.

When you observe two different rays AO and AM_1, we can come to the following conclusions.

1. Both are produced from Point A, but only the ray AM_1 entered the disk, so it has minimum energy.

2. The ray in the diagram disk to AM_1 is a tangent. The angle of OA M_1 is 90°. In this, ray OA is perpendicular to tangent AM_1. Hence, maximum energy and intensity will be present in ray AO.

3. Due to rotation, the ultra-cooled celestial body moves to position B after **t** time of interval. BO is the right-angle ray, and BM_1 is a slanted ray. But the slope of BM_1 is less than of AM_1. So in terms of energy, $BM_1 > AM_1$. It means ray BM_1 has more energy and intensity than AM_1.

4. In the same way, after time interval **t1**, the ultra-cooled astral body moves to position C from position B. Here, the slope of CM_1 is less than of BM_1. So in terms of energy, $CM_1 > BM_1$. This means ray CM_1 has more energy and intensity than BM_1. So in terms of energy, $CM_1 > BM_1 > AM_1$.

5. Similarly, after time interval **t5**, the ultra-cooled astral body moves to position F. Here, the slope of FM_1 is less. So in terms of energy, $FM_1 > EM_1 > DM_1 > CM_1 > BM_1$. It means ray FM_1 has more energy and intensity than previous positions.

6. Thus, the ultra-cooled celestial body changes its positions as follows:

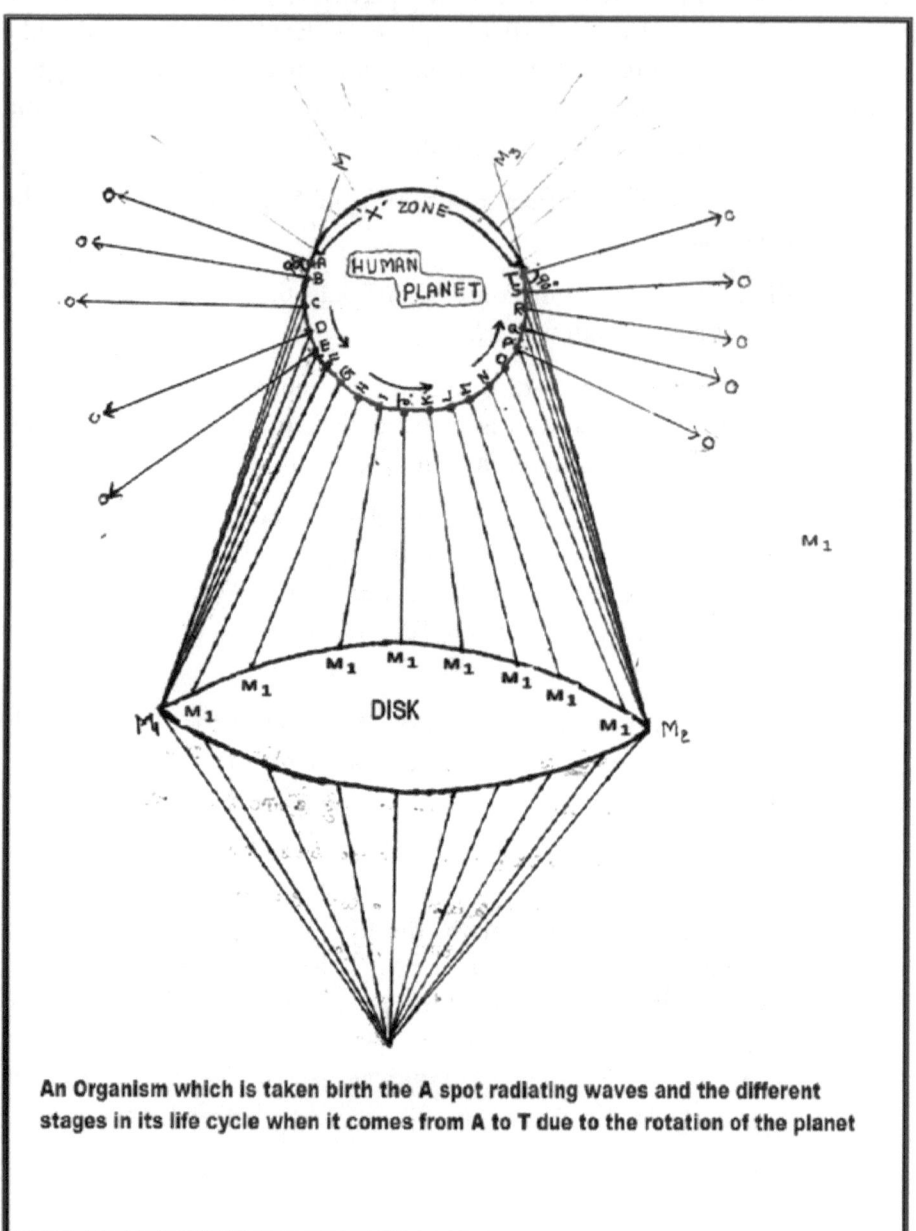

An Organism which is taken birth the A spot radiating waves and the different stages in its life cycle when it comes from A to T due to the rotation of the planet

GROWTH – RETENTION – DECREASE OF A CREATURE DUE TO RELATED PLANET ROTATION

S.No.	Time Interval	Position	Incidence of Ray	Energy Level
0.	T	A	High slope	+
1.	T1	B	Declining slope	++
2.	T2	C	Declining slope	+++
3.	T3	D	Declining slope	++++
4.	T4	E	Declining slope	+++++
5.	T5	F	Straight line	+++++
6.	T6	G	Straight line	+++++
7.	T7	H	Straight line	+++++
8.	T8	I	Straight line	+++++
9.	T9	J	Straight line	+++++
10.	T10	K	Straight line	+++++
11.	T11	L	Straight line	+++++
12.	T12	M	Straight line	+++++
13.	T13	N	Straight line	+++++
14.	T14	O	Straight line	+++++
15.	T15	P	Increasing slope	++++
16.	T16	Q	Increasing slope	+++
17.	T17	R	Increasing slope	++
18.	T18	S	Increasing slope	+
		T	High slope	

7. The amount of energy released from a point (bio-wave radiation point) of the ultra-cooled celestial body will be the constant during the time interval between t_4 and t_{14} because they are all straight lines, as they all pass through the disk. Hence, energy levels in this zone are the same.

$$GM^1 = HM^1 = IM^1 = JM^1 = KM^1 = LM^1 = MM^1 = NM^1 = OM^1.$$

8. The amount of energy released from the ultra-cooled celestial body will be the constant during the time interval between t_4 and t_{14},

because they are all straight lines as they all pass through the disk. Hence, energy levels in this zone are the same.

$FM_1 = GM^1 = HM^1 = IM^1 = JM^1 = KM^1 = LM^1 = MM^1 = NM^1 = OM^1 = OM^2$.

9. The amount of energy in the biowaves released from the ultra-cooled celestial body will be decreasing during the time interval between t_{14} and t_{18} because they are all sloped lines, as they are all not passing through the disk gradually. Hence, energy levels in this zone are decreasing.

$$OM^2 > PM^2 > QM^2 > RM^2 > SM^2 > TM^2$$

10. Ultimately, the amount of energy released from the ultra-cooled celestial body from point A to T can be presented as follows:

$AM_1 < BM_1 < CM_1 < DM_1 < EM_1 < FM_1 = GM^1 = HM^1 = IM^1 = JM^1 = KM^1 = LM^1 = MM^1 = NM^1 = OM^1 = OM^2 > PM^2 > QM^2 > RM^2 > SM^2 > TM^2$

11. From point A to F, the energy levels increase due to the reduction of the slope. From point F to O, the energy levels of biowaves are constant and maximum, as all of these are straight lines that pass through the disk. From point O to T, the energy levels are decreasing due to the increase in the slopes of the rays.

12. In the diagram, the ultra-cooled celestial body was shown rotating clockwise. Whatever may be the direction, the angles of incidence of rays do not change.

Relationship Between Energy of an Organism and Quality of Life

1. The number of cells in an organism is directly proportional to the amount of energy in the biowaves that an organism receives from the ultra-cooled astral body.

2. From point A, the amount of energy that is received from the ultra-cooled celestial body starts the life originating as a single cell.

3. As the energy increases, unicellular life will become multicellular, and complexity increases in organization and structure.

4. Thus, the amount of energy in the biowaves received from the ultra-cooled astral body is correlated with the various developmental stages of life as detailed below.

S. No.	Time Interval	Position	Incidence of Ray	Energy Level	Developmental Stage
0.	T	A	High slope	+	Single cell
1.	T1	B	Declining slope	++	Multicellular embryo
2.	T2	C	Declining slope	+++	Infancy
3.	T3	D	Declining slope	++++	Childhood
4.	T4	E	Declining slope	+++++	Adolescence
5.	T5	F	Strait line	+++++	Youth
6.	T6	G	Straight line	+++++	Youth
7.	T7	H	Straight line	+++++	Youth
8.	T8	I	Straight line	+++++	Youth
9.	T9	J	Straight line	+++++	Youth
10.	T10	K	Straight line	+++++	Youth
11.	T11	L	Straight line	+++++	Middle age
12.	T12	M	Straight line	+++++	Middle age
13.	T13	N	Straight line	+++++	Middle age
14.	T14	O	Straight line	+++++	Middle age
15.	T15	P	Increasing slope	++++	Old age
16.	T16	Q	Increasing slope	+++	Old age
17.	T17	R	Increasing slope	++	Old age
18.	T18	S	Increasing slope	+	Terminal stage
		T	High slope		Death

In the life cycle of any organism, there are three phases. They are:

1. Growth phase.

2. Stagnant phase.

3. Reduction phase.

If you take human life as an example, the longevity of man is considered 100 years. The various phases of human life with their periods are shown below:

S. No.	Position	Incidence of Ray	Developmental Stage	Time Period (Years)
1.	A to B, C, D, E, F	Declining Slope	Growth	20
2.	F to G, H, I, J, K	Straight Line	Stagnant youth Age	30
3.	K to L, M, N, O	Straight Line	Stagnant middle age	30
4.	P to Q, R	Increasing Slope	Old Age	20
5.	R to S, T		Terminal Stage	

The waves that are produced from points A to T can only influence the life. After point T, death occurs. As there is no possibility of waves entering through the disk that are produced from zone x, they cannot influence life on Earth.

ASTRO MATHEMATICAL DATA RELATED TO THE BIO-MASS OF A CREATURE

In the case of man, the time required from fertilization to growth is 20 years. In this period, the rays produced from the ultra-cooled astral bodies reduce from 90° to 0°.

The change in the angle of rays over 20 years from A to M = 90° to 0°

The change in the angle of rays over one year = 90°/20 = 4.5°

The Change in the Angle of Rays Over 10 Months

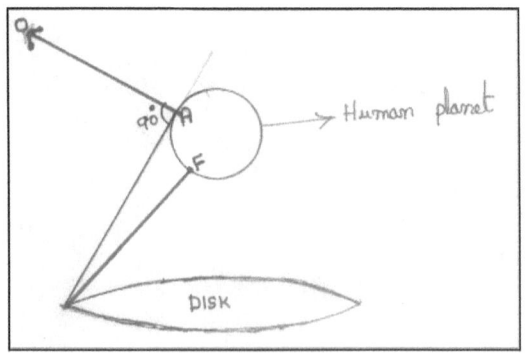

The pregnancy period = 4.5 × 10/12 = 3°45′

Then, the angle caused at point AM_1 through the disk = 90° − 3°.45′ = 86°.15′

At this stage in ray AO, the energy content of AM1 = AO/AM = Cos O = cos 86°.15

The value of cos 86°.15′ = 0.0645; Cos 0° = 1

According to biometry values supplied from Altra-sonography:

- Minimum just-born healthy baby weight = 2.5 kg
- Maximum = 3.5 kg
- Energy value of AM1 at the value of child birth = cos 86°.15′ = 0.0645
- Minimum amount of biomass that can be produced by a ray at an angle of 86°.15′ = 2.5 kg
- Maximum = 3.5 kg

The wave energy at 86°.15′ = cos 86°.16′ = 0.0645

In AM_1 inclination, the energy content of the right angle produced from the disk = cos 0° = 1

Cos 86°.15′ = 0.0645; Cos 0° = 1

Right angle biomass values are:

- If biomass values are minimum, 2.5 kg, then 2.5/0.0645 = 38.76 kg
- If biomass values are maximum 3.5 kg, then 3.5/0.0645 = 54.263 kg

From the above equations, we can deduce that at a right angle, the total biomass that can be produced is 38.76 to 54.263 kg. Here, the values of male and female weights are different. They are explained in part II and chapters 8, 9, and 12.

Question: Old people should receive emissions from the previous place of the ultra-cooled celestial body, which is in a right angle. This would prevent aging and should continue youth. But that is not happening. We should explore the reason for this.

Explanation

Even though many radio stations are airing their programmes, our radio can receive only one particular frequency of waves from the surroundings at a particular time and convert them into sound waves.

In the same way, while forming life with one particular frequency of waves releasing from a particular area, for example, A will influence the genetic material and DNA and respond to that frequency until the death of that life. But it cannot be influenced by any other type of wave frequency that exist in the surroundings.

The specific reasons for this are explained in part II and chapters 8, 9, and 14.

As there are plenty of rays produced from the surface area of the ultra-cooled celestial body, the number of living beings is also high, and they are all one living species, or races.

We already learnt that the longevity of an organism will be dependent on time, and the rotational velocity of the ultra-cooled celestial body.

There are plenty of ultra-cooled celestial bodies in the arm of spiral galaxy. The waves that are produced from these celestial bodies and are passing through the disk can only form different living being species or animal species. Similar changes are observed in the animal and plant kingdoms.

The details regarding the plants of the plant kingdom are explained in part III. The birth, growth, and death of lower organisms such as unicellular organisms are explained in part IV.

CHAPTER 2

Size of an Organism of a Species Responsible to Bio Waves Frequency of Related High Cooled Celestial Body

FORMATION OF BODIES OF CREATURES DUE TO BIO-WAVES EMITTED BY CELESTIAL BODIES

Different species of organisms are of different weights and sizes. The reason for this is that the amount of energy received from the different ultra-cooled celestial bodies through bio waves is different.

Normally, one ultra-cooled celestial body will produce one type of bio-wave, which can influence one particular species, and they will all have similar weight and size. Similarly, different ultra-cooled celestial bodies will produce different frequencies of bio-waves with different energy levels and influence different species. So different species will have different weights and sizes. Hence, smaller ultra-cooled celestial bodies will produce smaller volume species, and larger ultra-cooled celestial bodies will produce larger volume species.

We should scientifically explain the relationship between the volume and mass of a species and the size of the ultra-cooled celestial body.

If you take two objects of the same mass, size, and shape and leave them from the Earth and the moon with an escape velocity of their gravitational force by using rackets and calculate momentum, they will have different momentum, even though the mass is the same.

Object mass = M

Momentum = Mass × velocity

Escape velocity on Earth = 11.5 km/s

So momentum on Earth = M × 11.5 km/s

Object mass = M

Escape velocity on Moon = 2.5 km/s

So momentum on Moon = M × 2.5 km/s

So momentum varies with different sizes of planets with the same mass.

Quanta

Pockets of energy are called quanta. The energy pockets of light are called photons. Energy pockets of sound are called phonons. The energy pockets of gravitational waves are called gravitons. Similarly, the energy pockets present in bio-waves can be called biotrons.

In order to know the energy content of different biotrons received from different ultra-cooled celestial bodies, we should calculate the momentum of a biotron.

The energy of a wave in a biotron = $E = h\nu$

E = Energy, H = Planck constant, V = Frequency, λ = wave length

$V = c/\lambda$

$E = hc/\lambda$

According to Einstein theory $E = mc^2$

$m = E/c^2$

But $E = hc/\lambda$

Then $m = hc/\lambda c^2 = h/hc$

Then momentum of a biotron = $mc = h \times c/\lambda \times c = h/\lambda$

According to the calculations of the momentum, the cell is linked to wavelength.

The escape velocity of an article depends on the mass of the body (Escape velocity = $\sqrt{2GM}$). Small planets will have less mass and large planets will have more mass, so the escape velocity of larger planets is higher than of smaller planets.

The bio-waves that are produced from small ultra-cooled celestial bodies will have less momentum and higher wavelength. If wavelength increases, the energy content of that biotron decreases. This leads to the decrease in frequency and number of biotrons produced, and ultimately leads to decreased size.

Similarly, the bio-waves produced from larger ultra-cooled celestial bodies will have more momentum and lower wavelength. If wavelength decreases, the energy content of that biotron will increase. This leads to the increase in frequency and number of biotrons produced, and ultimately leads to increased size.

In the family of ultra-cooled celestial bodies, those at the axis have more weight than others. For example, the ultra-cooled massive celestial body (nonluminous Star Ex: our Solar system) weight is 99.86% and the rest of the planetary bodies and satellites in their family weigh 0.14%.

Previously, we proposed that one nonluminous star gets one plant species, one planet to one herbivore, and one satellite to one carnivore. Here, nonluminous stars are more massive than any other. So plants will be of higher sizes. The plant biomass on the Earth is higher than the mass of animals and other living beings. Herbivores are smaller in size than plants. The weight of the biggest herbivore, the elephant, is less than a medium-sized plant.

All plants weight should be more than animals, but some perennial plants and grass varieties weigh less than animals. In the same way, certain algae such as spirogyra and euglena have lower weight than animals. The reasons for such anomalies are explained in detail in part III.

According to the one planet-one animal hypothesis, the total biomass of animals should be less than the total biomass of plants. This is true in the case of Earth. The total number of planets is greater than the number of nonluminous stars. In the same way, the total number of animal species is more than the total number of plant species. The total number of plants on the Earth is around 4,00,000 and the total numbers of animals is 15,00,000.

The total number of carnivores is more in the animal kingdom than herbivores. Animals such as ox, buffalo, zebra, rhinoceros, and elephant are bigger, and they are herbivores. The planets relating to these animals may be massive.

The one satellite-one carnivore concept states that satellites are smaller than planets, hence the size of carnivorous animals will be smaller than herbivorous animals. There are certain exceptions to this. Some carnivores, such as tigers, lions, and wolves, are bigger than other animals, even though they are carnivorous. The special circumstances for the occurrence of such animals are explained in part II, chapter 13.

On Earth, there are some animals that are smaller. For example, hydra, paramecium, euglena, bacteria, virus, etc. The ultra-cooled celestial bodies that are responsible for the genesis of these organisms should be smaller.

Previously we studied that Fungus, Bacteria and Virus species generating celestial bodies emitted bio-waves are reaching the earth directly and not through the Disk. Disk facilitate the living beings to under go the different stages of life as we studied previously from germplasm, embrio, infancy, childhood, adolescence, youth, adult, oldage and death. The fungus, bacteria, virus, euglena and paramecium such as living beings generating celestial bodies are even though they large, their emitted waves are not reaching the earth through the Disk. Hence they cannot show the different stages of life in their life span.

CHAPTER 3
Sexuality in Plants, Animals, and Lower Organisms

The gravitational pull determines the sex in the animal creatures. There is wide diversity in the sexes of organisms. Some are monosexual, some are bisexual, and some are sterile. In monosexual organisms, males and females are separate. This is called sexual dimorphism. In certain organisms, male and female parts are present in the same organisms. Such animals are called hermaphrodites.

Sexual dimorphism is more present in herbivores and carnivores than plants.

AREAS OF HIGH COOLED CELESTIAL BODY (PLANET) THAT PRODUCE OF MALE, FEMALE AND NEUTER GENDERS

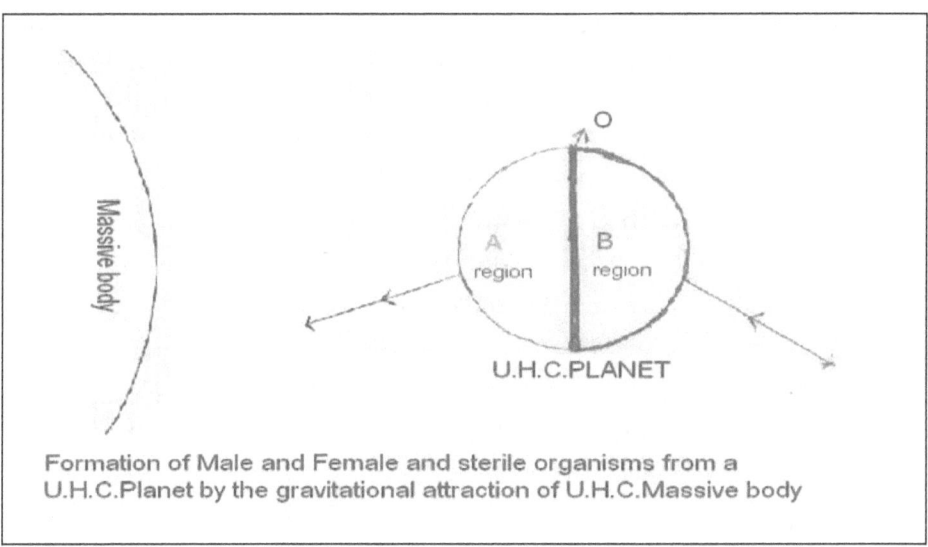

Formation of Male and Female and sterile organisms from a U.H.C.Planet by the gravitational attraction of U.H.C.Massive body

Reasons for sexuality in animals: Process of body formation in males and females by bio-waves.

1. In the family of ultra-cooled celestial bodies, the celestial body (non-luminescent star) that are at the axis have more weight and are more massive. Hence, here it is being called as a massive body.

2. Some planets and satellites revolve around these massive nonluminous stars.

3. Due to the influence of the gravitational force (waves influenced by massive body's gravitational force and it is proved by Einstein theories) of the massive nonluminous star, on the bio waves emitted by the planet, the entire area on the ultra-cooled celestial planet will be divided into two zones as shown in the figure.

4. Zone A will have positive attraction force by the massive body, as it was facing the massive body, so the bio-waves produced from this area will have more frequency (because zone A area of planet released waves are towards of massive body) than real, producing more mass in the animal, leading to the birth of male organisms on Earth.

5. Zone B will have negative attraction force (due to back side of the planet) on the bio-waves produced from this will be effected and its frequency will be little less than real, (planet B side emitting bio waves are attracting towards massive body side so their frequency will be little less than real) so they produce less mass in the animal, leading to the birth of female organisms on Earth. (This is why male organisms are a little bigger than females.)

6. Zone O, where the Zone A and Zone B merge, will produce sterile animals. The area of this region is very low, so the number of animals that are sterile in animal species is in lower quantity when compared to the quantity of both male, female totally.

7. This above principle will be true when we observe all species in Animalia, as every planet divides into A and B regions by massive body gravitational attraction force.

8. There are certain exceptions to the above principle of sexuality, however, even though the majority of males are bigger than females in the animal kingdom. In animals belonging to the phylum *Nemati*

helminthes, females are bigger than males This can be explained in the following way:

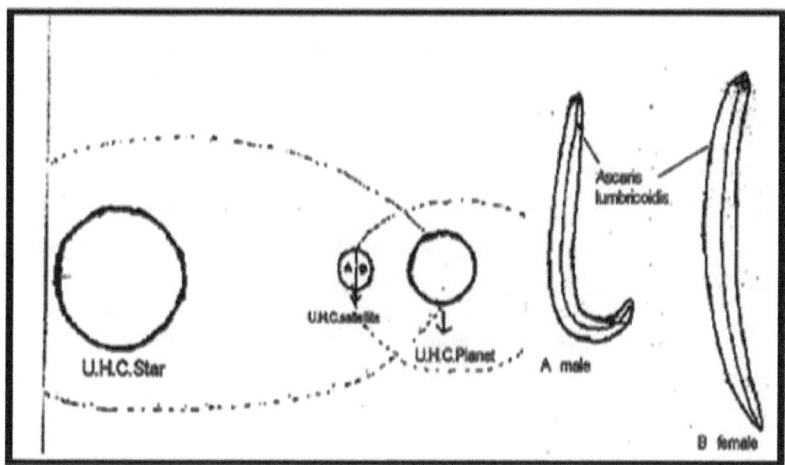

9. In the case of satellites moving around ultra-cooled massive stars, they are also influenced by gravitational forces of the planets in that system. This means these satellites are not only influenced by massive stars but also by planets. In such cases, the planets are nearer to satellites than the massive stars.

10. As a result, the influence of the planet on Zone B is higher than the influence of the massive star on Zone A. Hence, Zone A produces smaller males and Zone B produces larger females.

In the animal kingdom, larger animals are present in phylum Chordata. So animals of the phylum Chordata are controlled by planets. So in almost all organisms of this phylum, males are bigger than females, and a few are sterile.

THE FORMATION OF NON-GENDER VARIATION IN PLANTS BY BIO-WAVES

Sexual Dimorphism in Plants

1. Plants receive energy from ultra-cooled massive non-luminescent stars. The bio-waves that are produced from the ultra-cooled massive non-luminescent stars are of the same type and will have same momentum (not divided into zone A and zone B). Because they are

not influenced by surrounding planets or satellites (itself as a massive body) and far away to the surrounding stars, the majority of the plants in the plant kingdom are bisexual.

2. Only at the reproductive stage of plants male and female parts such as pollen grains and ovary are developed in flowers. Only few plants are monosexual . This can be explained in this way.

3. Sometimes, ultra-cooled massive non-luminescent stars are influenced by the surrounding ultra-cooled massive non-luminescent stars in such cases the area on the ultra-cooled massive non-luminescent stars is divided into two zones as shown in the diagram.

4. Zone A will have more attraction force and thus will produce male plants, and Zone B will produce female plants.

5. The centre of the galaxy is far from the ultra-cooled massive non-luminescent stars, so its influence on them is lower under normal circumstances.

6. At some times, two ultra-cooled massive non-luminescent stars will pair with each other and form a system. In the Milky Way galaxy, 20% of them are this type. Here, ultra-cooled massive non-luminescent stars are also divided into two zones, as shown in the diagram.

7. Zone A will have more attraction force and thus will produce male plants, and Zone B will produce female plants by negative attraction force.

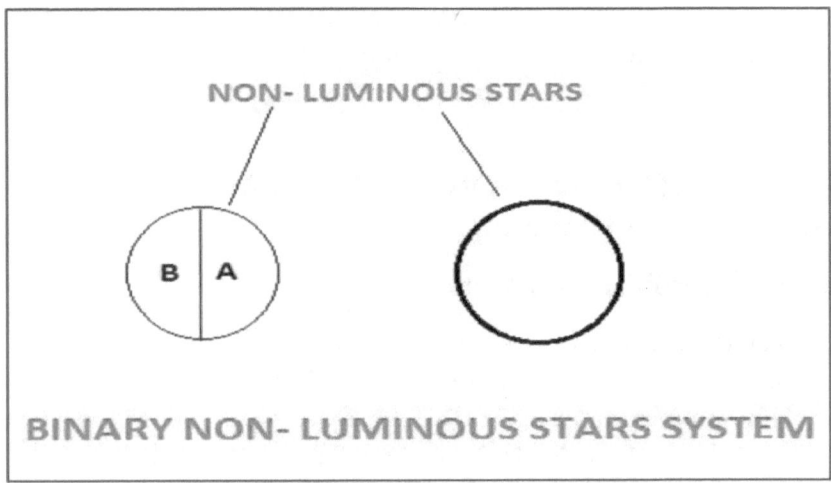

8. The reasons for sexual diversity at a particular phase of the growth of plants, such as production of pollen and the ovary, are explained in part III.

Reasons for Asexual Reproduction in Lower Organisms

1. Certain lower organisms like paramecium and amoeba experience asexual reproduction.

2. AXIOM 1, whose bio waves, emitted by the nonluminous star families, passes through the DISK and centralised on earth will only creating the plants, herbivore and carnivore species organisms so with the help of the DISK these organisms exhibiting the different stages like single cell, embryo, infancy, childhood, adolescence, youth, middle age, old age and to death in their life spans.

   AXIOM 2, whose bio waves are emitted by the nonluminous star families and DO NOT PASS through the DISK but fall directly on the earth, create various fungus, bacteria and virus species organisms on earth but organisms of these species do not exhibit variations like said the aforesaid, in axiom 1, due to the missing help of the DISK. Hence, these organisms will mainly follow asexual reproduction like binary fission, multiple fission, budding etc. Sometimes, these organisms also follow sexual reproduction like male, female gametes. These are all separately explained clearly in PART 1V- ASTROMICROBIOLOGY.

3. The reasons for producing male sexual organs in animals of Zone A and female sexual organs in Zone B are explained in part II, chapter 8.

4. Similarly, reasons for sexual variability such as XX male, XY female in the case of certain birds and moths, XO male XX female in the case of grasshoppers and bedbugs, ZZ males, and non-disjunction and conditions causing AAXXY, AAOY, and AAXXX are explained in later chapters.

CHAPTER 4
Biofield Mode of Non-Segment and Segment Structure Animals By Biofield Type 1 and Type 2

In this chapter, the influence of the biofield in determining the specific shape of the organisms is discussed. The knowledge of physics states that the life that receives charged particles or charged waves will produce charged fields. We know that the earth has a magnetic field. Similarly, every ultra cooled celestial body (either star, planet or satellite) has its own protomagnetic field, due to producing protomagnetic waves.

The concept of a bio field is explained here below:

When you keep a bar magnet in the Earth's magnetic field, it produces two different types of magnetic fields. Similarly, every high cooled planet or satellite is in the nonluminous star field. Here, nonluminous star family is an object. Its image is focused on earth in the form of living-being field. This means the field of a bar magnet in the earth magnet field is correlative with the living being bio field. The similarities are given clearly below.

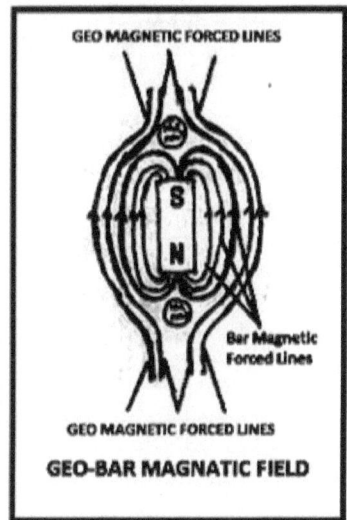

GEO-BAR MAGNATIC FIELD

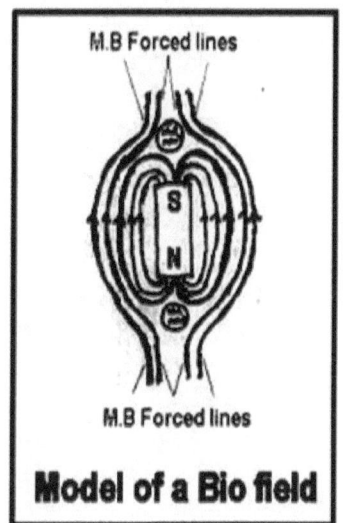

Model of a Bio field

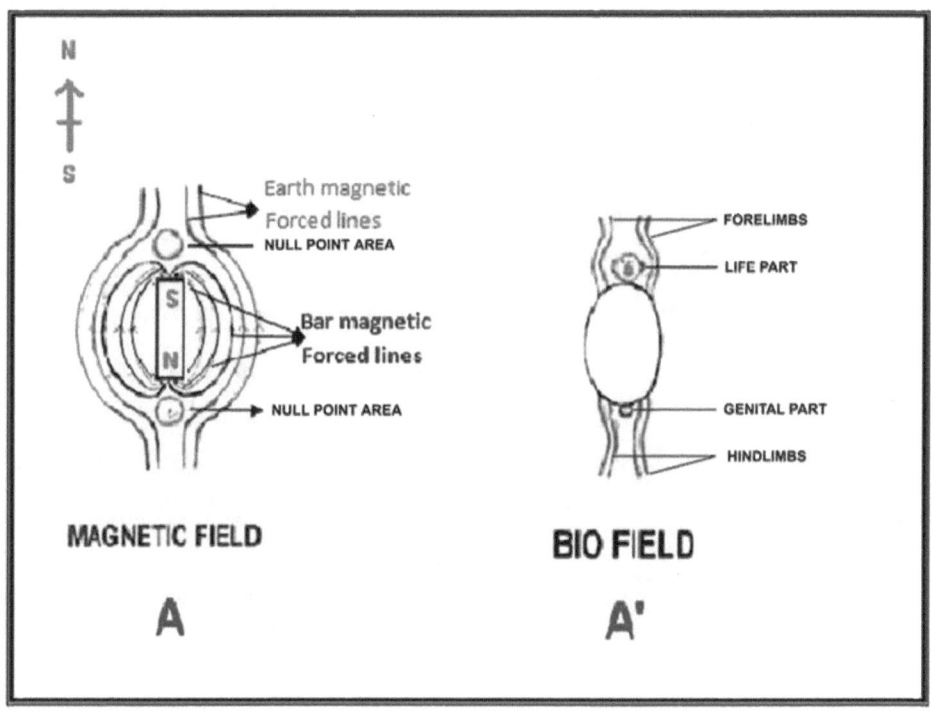

MAGNETIC FIELD
A

BIO FIELD
A'

S.No.	Diagram A	Diagram A^1
1.	Field of bar magnet.	Can be compared to the trunk of an organism.
2.	Above the geomagnetic force lines, an upper null point area is present.	Above the massive body magnetic force lines, forelimbs are present on both sides of the upper null point area.
3.	The force lines influenced by the bar magnet are present on both sides of the lower null point area.	At the geomagnetic force lines, hind limbs are present on both sides of the lower null point area.
4.	Just below and above the bar magnet, one null point is present.	Just above the bar magnet at one null point, the head is present. Genital organs are present at the null point just below the bar magnet.
5.	The bar magnetic force lines are arranged like scales in an onion. Only the outer layer can be seen directly.	In the same way, we can only see the outer layer of the trunk.

In the above diagrams, there are similarities between diagram A and A^1. They are listed in the table:

The following variations are also found in the above diagrams:

1. Normally, in any organism, the forelimbs are not projected towards the upper side, as shown in the diagrams.

2. There are no force lines at the null point. There is no possibility of forming any mass at null points. But the head is present at the null point, which does have some mass.

3. The area at the null point is formed due to opposite forces meeting, thus making it neutral. So there is no possibility of forming any organs here. In contrast to this, we find the eyes, ears, nose, and mouth in this area.

4. Even though massive force lines are present surrounding the trunk, only forelimbs and hind limbs are formed.

Explanation

1. In any organism, the forelimbs are not projected towards the upper side as shown in the diagrams, even though the force lines face the upper side. If the forelimbs are oriented to the upper side, the organism has to spend a certain amount of energy to counteract the geo gravitational force acting on it. In order to conserve energy, they are adapted to the downward direction of the forelimbs. As the forelimbs are oriented against the field force lines, the amount of energy also decreases in the forelimbs. To compensate for this, we occasionally move and stretch our forelimbs towards the upper side. This is something like yawning. When our body feels the presence of less bio energy, our body yawns and stretches our fore and hind limbs according to the bio field and improves our bio energy. Ultimately, we gain sufficient biotic energy. This can be observed not only in human beings but also in other animals.

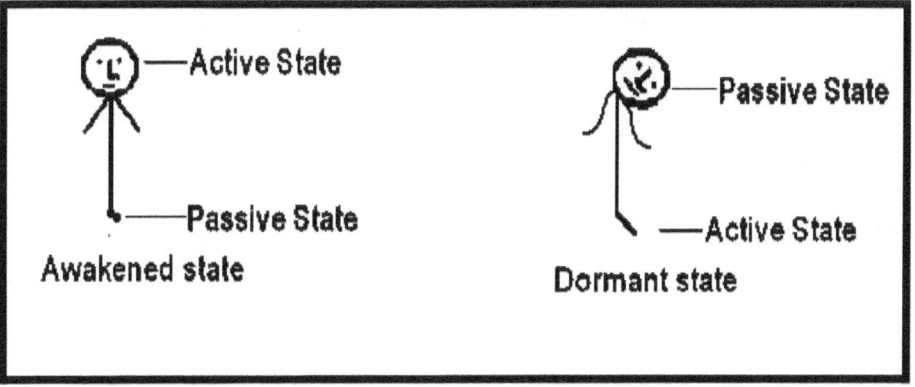

2. In any organism, there are two states. One is the awakened state and other is the dormant state. In the dormant state, the head and limbs are slacked and in a relaxed position. As if they are not related to the individual. But in the awakened state, they change their position by stretching their limbs and head according to the bio field force lines. At null points, there are no magnetic forces. But the head of an individual is formed here because a bar magnet is stable and lifeless. But in the case of an organism in the dormant state, the head and limbs are slacked and in a relaxed position, as if they are not related to the individual. But in the awakened state, they change their position

by stretching their limbs and head according to the bio field force lines. That is why in the awakened state, due to upward mobility of bio force lines, the null point changes and forms above the head. At this time, at the lower end point, the force lines and field force is less.

3. In the awakened state, the genital organs in the lower end are slacked as they are actually in a null point area, and they are in a relaxed position as if they are not related to the individual. But in the resting state, there is the influence of force lines in the lower end, and reproductive organs are active in the resting state. This is true both for human beings and for animals.

4. The reasons for the formation of the head, eyes, ears, nose, mouth, and genital organs at the null points are explained in part II, chapter 6.

5. Even though massive force lines are present surrounding the entire trunk region, only two upper and two lower limbs are formed, instead of many limbs. The reasons for this are explained in part II, chapters 5 and 8.

Studies on the influence of a bar magnet in a geomagnetic field.

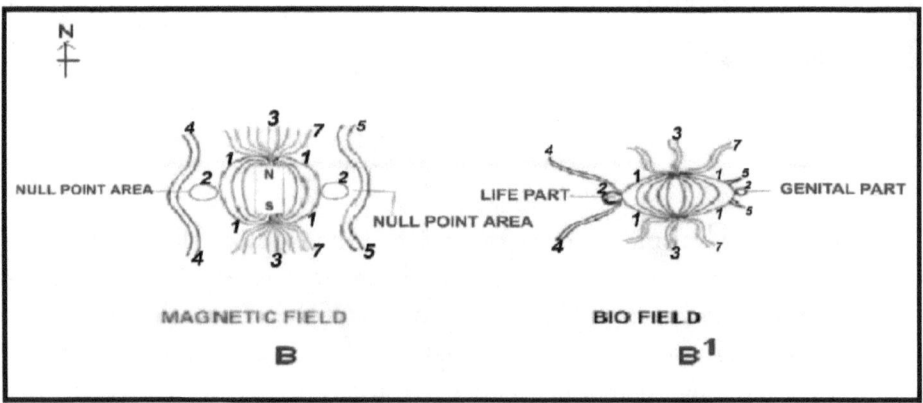

In the above discussion, we have studied one type of field orientation. But when we study the orientation from diagrams B and B^1, we can come to the following conclusions:

In the above diagrams, there are similarities between diagram B and B^1. They are listed in the table below:

S.No.	Diagram B	Diagram B^1
1.	Force lines touch the poles of bar magnets.	These force lines form the main part of the body.
2.	Perpendicular to the bar magnet, two null point areas are present.	At one null point area is the head, and at the other null point area the genital organs form.
3.	From north and south poles of the bar magnet, numerous geomagnetic force lines appear.	Due to these force lines, plenty of limbs are formed.
4.	Adjacent to left side null point, many geomagnetic force lines are present.	These force lines are formed into antennae.
5.	Adjacent to the right side null point, many geomagnetic force lines are present.	These force lines are form anal organs.
6.	Many force lines are present near each other in a bar magnetic field.	These force lines are responsible for segmentation, thus forming segmented living beings.
7.	At the north and south poles, more geomagnetic force lines are present.	These magnetic lines form limbs, and thus six to eight limbs form in an animal.

Segmented Animals

From the diagrams A and A1, we have learnt that due to the orientation of N → S, S → N field forces, non-segmented four-limbed organisms (two forelimbs and two hindlimbs) are formed. In the same way, from diagrams B and B^1, we have learnt that due to the orientation of N → S, N → S field forces, segmented multi-limbed organisms (tetrapods, hexapods, octopods, and decapods) are formed.

The majority of segmented animals are present in Phylum Arthropoda. Out of the total 15,00,000 Animal species, 12,00,000 are segmented Animal species. The ant is a small, segmented animal. There are six legs for its movement. Are this many legs required for its movement? A cockroach has six legs in addition to two wings. Are all of these needed for its movement? The reason for all of these is the biofield and influence of magnetic field orientation in the N → S, N → S direction.

Even though there are only two magnetic field orientations:

1. Segmented limbless animals, e.g. Annelids, earthworms.

2. Non-segmented limbless animals, e.g. Reptiles, snakes.

3. Segmented multi-legged animals, e.g. Annelids, centipedes, and millipedes.

4. Segmented and presence of limbs in larval stage only, e.g. Arthropods, Lepas.

5. Animals without a definite shape, e.g. Molluscs and echinodermates.

The reasons for the above categories of organisms are explained in part-II, chapters 5, 12, and 13.

In the same way, the shape of plants and their structures are explained in part-III Astrobotany.

From this chapter, it can be inferred that the influence of the magnetic field orientation in the N → S, S → N direction in the biofield is responsible for different shapes of organisms.

CHAPTER 5
Shape and Size of an Organism

EFFECT OF MASSIVE BODY (NON-LUMINOUS STAR) ON A PLANET

In the solar system, small planets like Mars, Venus, Earth, and Mercury are formed nearer to the Sun; large planets like Jupiter, Saturn, and Uranus are in the middle; and planets like Neptune and Pluto are at the end.

Model of Non-Luminous Solar System

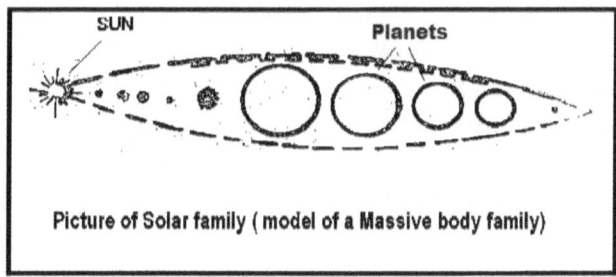

Picture of Solar family (model of a Massive body family)

ANIMAL FORMATION AND BIRTH

In the same way, the ultra-cooled nonluminous massive body family will also have smaller planets nearer to it, bigger planets in the middle zone, and smaller planets at the other end, with meteors and Oort clouds present further away from this ultra-cooled non-luminescent massive star family. Previously we discussed that every non-luminescent star and its planets and satellites are having protomagnetic field around themselves. According to Coulombs law in Physics states that $F = K \frac{q_1 \times q_2}{d^2}$. Here is F = electrical force, q_1 = charged particle, q_2 = another charged particle, d = distance. According to this law electrical force between two charged objects (charged celestial bodies) is directly proportional to the product of the quantity of charge on the objects and inversely proportional to the square of the separation distance between two objects. K is the Coulombs law constant. Now we can come to a concept that is, in the above diagram star (non-luminescent star) has a protomagnetic field and its field covers entire its family celestial bodies. The energy in the forced lines of the star, near is high and low far. This means taking as model of our solar system, the force in the forced lines at the Mercury planet place is high than the Venus place and also decrease gradually according to the distance from the Sun increase. From this, we can conclude that the influence on the bio field produced from planets nearer to the massive body is higher than the biofields produced from planets further away.

This means the magnetic force influence on the planets is inverse squire proportional to the distance between the planet and the nonluminous massive celestial body. Thus, the variations in shape, size, habit, and habitats of animals are due to different biofields that are produced by different planets and their distance to the massive nonluminous star.

Changes in Animal Species Due to Orbits of Ultra-High-Cooled Planets

The majority of the large animals are present in the chordata of the animal kingdom. There are around 45,000 species. The rest of the phyla has smaller animals. We already learnt in Part II that the larger animals are formed by planets, and not by satellites or small planetary particles. Thus, we can conclude that larger animals are related to planets. But we have to know which planet is producing which species.

For the sake of our convenience, we can divide Chordata into Pisces, Amphibians, Reptiles, Aves, and Tetrapods. The genesis of these different animals from different planetary bodies is explained below:

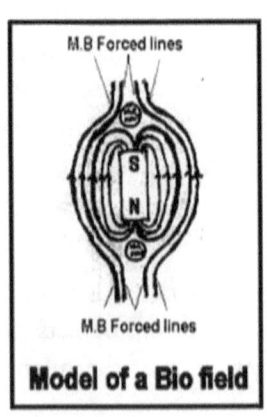

Model of a Bio field

1. **Aves:** Birds are formed from the biofields produced by planets that are nearer to massive nonluminous star bodies. These massive nonluminous star bodies produce strong force lines in the upward direction. The geomagnetic force lines (This terminology is only being used to help understand. Really, the words used differ and they will be explained in further chapters) that are towards the north side produce wings. In the same way, the geomagnetic force lines towards the south side produce hind limbs, or legs. Normally, the area of the null point is triangular. So in the case of birds, a triangular, sharp, protruding beak can be observed. In the family of ultra-cooled nonluminous massive star bodies, smaller planets are nearer to them, bigger planets are in the middle zone, and smaller planets are at the other end, and Oort clouds are present far away. These smaller planets will produce small birds, and larger planets will produce larger birds. For example, first small birds like doves, crows, and sparrows (carinata class) are formed from smaller planets. Next to these, large birds are formed from bigger planets, so these birds are bigger. As they are far from the massive body, the force lines that are produced in the biofield are not as strong as in the case of the carinata class. So flightless birds such as the ostrich and the kiwi (class Ratite) are formed. Related figure is shown below:

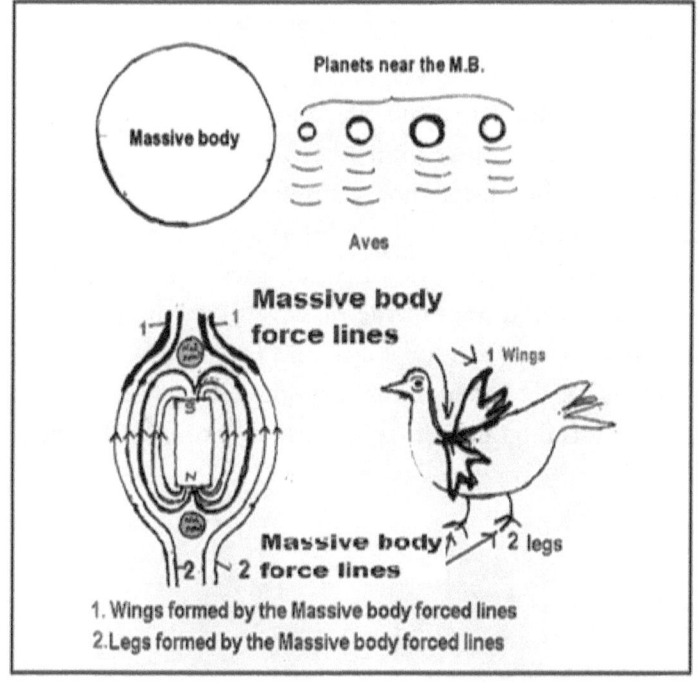

2. **Tetrapods:** Tetrapods are formed from planets that are neither big nor small. The force lines for these animals are neither strong nor weak due to some more distance to the massive star, so they lost the ability of flight and instead move on Earth and have adapted to terrestrial life. The areas of the null points are triangular, so their head shape is also triangular. In the case of large planets, the force lines are stronger, so large animals like cattle, elephant, and rhinoceros are formed. Related figure is shown below:

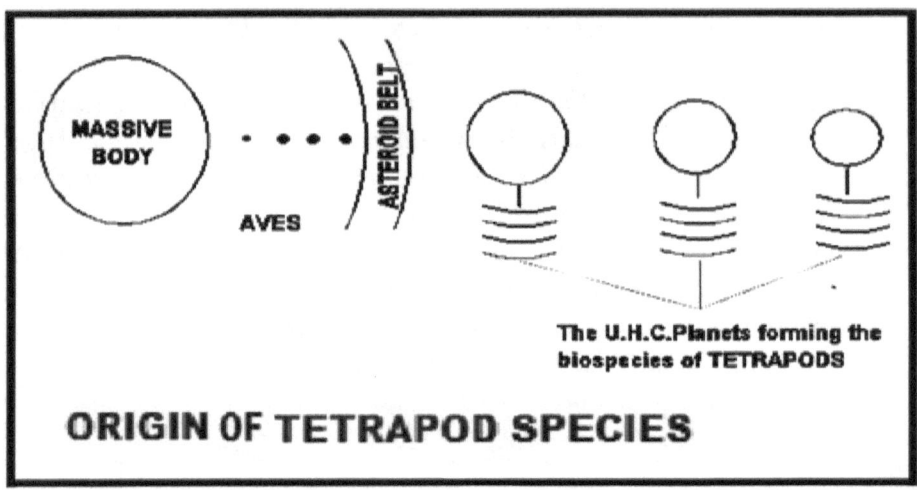

3. **Pisces:** The planets that are far from the ultra-cooled non-luminous massive star body cause biofields, which form Pisces. The magnetic force lines of the nl.star (non-luminous star) in the place of far away planets are weak, and their limbs, which are formed by these force lines, are weak. In the case of fishes, these become their fins. As the force in the forced lines of the massive nl.star body is lower, they are not related to any plant on Earth and live independently in the water by feeding algae. In the same way, the area of the null point is also low, thus forming animals with little or no neck. In the majority of

fishes, the head is merged with the trunk, so there is no neck region. Related figure is shown below:

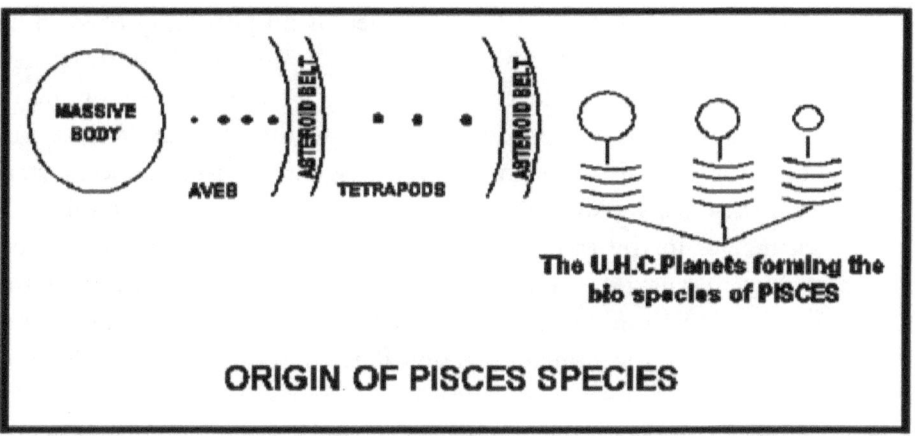

ORIGIN OF PISCES SPECIES

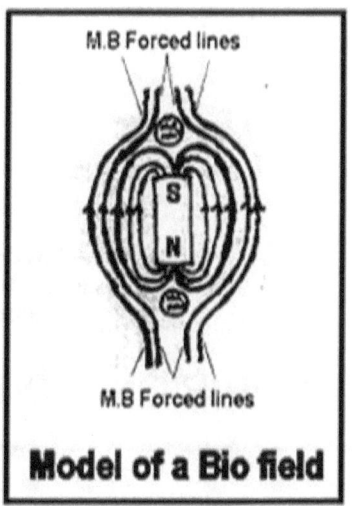

Model of a Bio field

4. **Amphibians:** We can find an asteroid belt between Uranus and Neptune or Jupiter and Mars. There are many asteroids. Such asteroids are responsible for the production of amphibians. Because these asteroids have no specific orbit so they belong to both Uranus and Neptune orbits hence these asteroids produced biofields exhibits both characters so they can live either on Earth or in the water. They have characteristics of both tetrapods and fish. Asteroids are smaller

than the planets, so animals that formed from the biofield of these asteroids should be smaller. This is confirmed by the zoological studies as evidence that smaller animals of the amphibian classes gymnophiona, urodele, and anura. There is no possibility of forming the first asteroid belt (between Mars and Jupiter) in the ultra-cooled nonluminous massive star body family, because massive luminous star bodies (stars) will produce heat due to thermonuclear reactions. This heat will distort planetary motion, but non-luminescent stars are themselves cool,So there is no possibility of the first asteroid belt existing, as in the case of solar system. Due to differential rotation of stars in galaxies, there will be changes in the outer orbits of the planets, thus forming the asteroid belt. That is why we are able to see amphibians in the animal kingdom. Related figure is shown below:

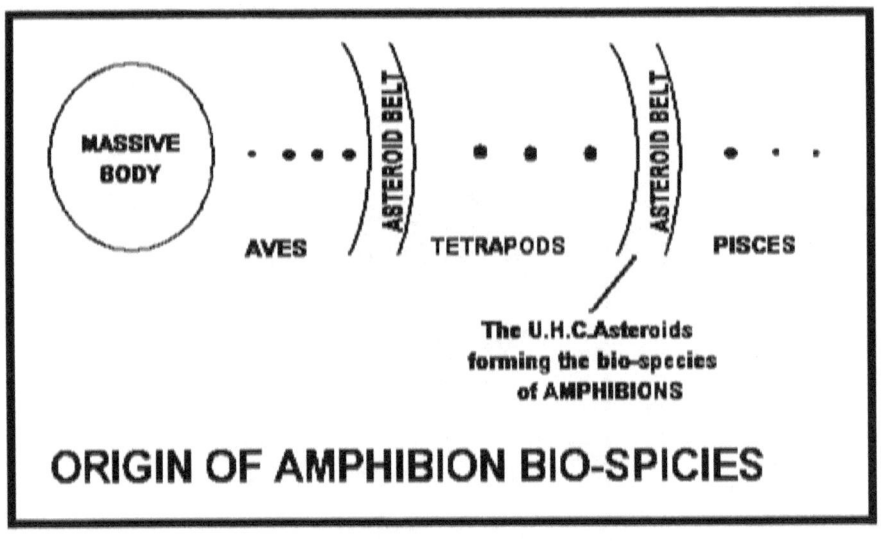

ORIGIN OF AMPHIBION BIO-SPICIES

5. **Reptiles without Limbs:** In the solar system, comets are sometimes formed from Oort clouds. This was confirmed by famous scientist Oort. A similar phenomenon may happen in the case of an ultra-cooled nonluminous massive star family forming limbless animals called reptiles from comet like structures. The force lines of the comets or Oort clouds are very weak because they are far from the nonluminous massive astral body, so they cannot form any limbs, as in the case of snakes. As the force lines are lengthy and hyperbolic,

the animals are cylindrical. There is no neck in these animals. The number of such animals is low. This is also in consonance with the small number animals (3,000 species of snakes) of the order squamate of class ophidia. Related diagram is shown below:

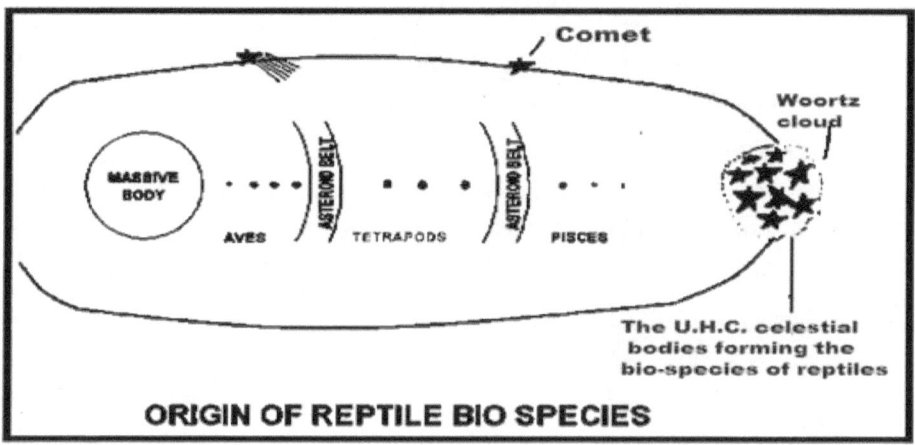

ORIGIN OF REPTILE BIO SPECIES

We Can Find the Following Additional Information from This Chapter

1. Celestial bodies have two types of rotation: rotation around themselves and rotation around other astral bodies.

2. The production of biotrons due to the inclination of rays; the growth, increase, decrease, and cessation of intensity of waves; and birth, growth, senescence, and death were explained earlier.

3. The nature of rotation of one celestial body around another astral body is not similar at all times. If the rotating celestial body is nearer to its influencing celestial body, it moves in a round orbit. If it is in medium distance, it rotates in an elliptical orbit. If it is at long distance, it revolves in a hyperbolic orbit. If it is very far away, it rotates in an ultra-hyperbolic orbit.

4. The masses of all celestial bodies are not same.

5. Sometimes, one planet comes under the influence of another planet.

6. Sometimes, one galaxy comes under the influence of another galaxy.

7. Sometimes, one star comes under the influence of another star, and they move as a pair.

8. There will be two states in motion of a celestial body around related massive body. At a particular time, celestial body will be very close to related massive body. This state is called epihelion.

9. At a particular moment, the celestial body will be very distant from the massive body. This state is called perihelion.

All birds are smaller than tetrapods because birds were formed due to biofields from smaller planets. There is no exception to this. You cannot find anything the size of an elephant.

In a similar way, Pisces are smaller than the tetrapods. Here, Pisces are formed due to biofields from distant small planets. There is one exception for this: whales. The reason for the genesis of whales with larger size is explained below.

Due to variable rotation, it is rare for a non luminous star body comes under the influence of another non luminous massive star body. At that time, one ultra-cooled non luminous star body will come under the influence of another ultra-cooled nonluminous massive star body. In such a case, the ultra-cooled non luminous star body that is influenced by another ultra-cooled nonluminous massive star body, loses its massive body nature and becomes a planet that influences the massive body. At this point, it stops producing plant species and begins to instead produce an Animalia species. Since it was large, it will produce large organisms, and since it was rotating in the outer orbit, the produced Animalia species also becomes Pisces. That is how whales are formed.

Due to differential planetary motions at epihelion or perihelion, different types of species are to be produced from the same ultra-cooled celestial body. But this was not observed in the animal kingdom. Otherwise, these planetary bodies should be in static orbits, as in the case of an electron moving in an atom (Neil Bohr's model). According to Neil Bohr's model atoms of the matter having orbits around the nucleus. These orbits named K, L, M, N are having limited energies. This means orbits are quantised.

Ultimately, the variations in patterns of life are explained through the influence of biotrons on genetic material that is present in chromosomes of the nucleus in chapters 9, 10, and 11 of part II.

CHAPTER 6
Method of Formation of Important Organs Like Eye, Ear, Nose, and Mouth by Fundamental Forces

FORMATION OF EYES, NOSE, MOUTH AND EARS BY FOUR FUNDAMENTAL FORCES

All the matter in the universe is made of atoms. These atoms are made up of atomic particles like protons, neutrons, and electrons. All these atomic particles are influenced (except Leptons) by the following four different natural forces:

1. Electromagnetic force.
2. Gravitational force.
3. Strong nuclear force.
4. Weak nuclear force.

Life is also formed by atoms. So life is also under the influence of the above four forces. That means, life should also respond to the above four types of stimuli differently. For that, we will take the example of an animal and study the force fields and their influence on life.

From the diagram, we can infer that surrounding the bar magnetic force lines, geo magnetic force lines (word used only for understanding) are present. It means that bar magnetic force lines are embedded in the geo magnetic force lines.

The matter or cells in the biofield cannot respond differently for the four different natural forces because the massive body force lines such as geo magnetic force lines are surrounding them. In order to get the influence of these four different natural forces, there should be some suitable place to do so.

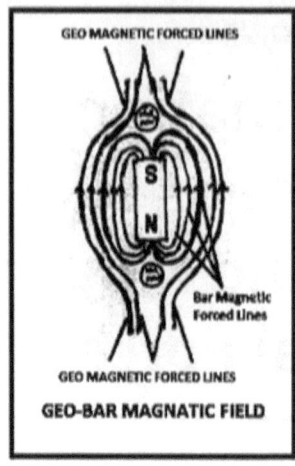

GEO-BAR MAGNATIC FIELD

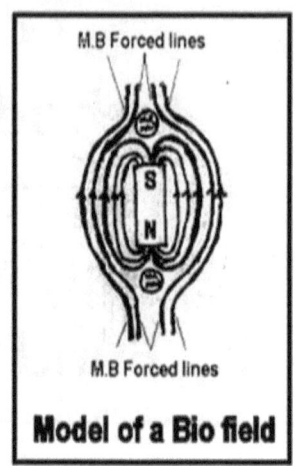

Model of a Bio field

From the above diagram, we can notice that the influence of the massive celestial bodies will be neutralized by the four forces only at the north and south poles. We call these points as null points. As these forces counteract the magnetic forces in the upward direction, four important organs are formed in that area. Now we have to know which type of force forms which type of organ.

1. **Electromagnetic Force:** Electromagnetic waves carry electromagnetic forces. The eyes are formed in the head as a response to receiving these electromagnetic forces.

2. **Gravitational Force:** All matter in the universe is under the influence of two gravitational forces: self-gravitational force, and gravitational force that can be exerted by anything. Normally, a smaller body will be attracted by a larger body and will try to move from its axis towards the larger body. To counteract this force, the smaller body will also use its self-gravitational force.

    a. The size of celestial bodies is increasing due to this law of universal gravitational force by adding smaller matter to the larger mass bodies.

    b. In the case of the biofield of an organism, there is only an upward self-gravitational force at the null points, which is responsible for the formation of matter in the form of the mouth.

    c. Even though we know there are attraction forces between two masses, the reason for such universal gravitational force is not understood so far. Famous scientist Albert Einstein tried to explain this phenomenon in terms of space and time, but he could not give the exact reason for this universal gravitational force.

3. **Strong Nuclear Forces:** These forces stabilize the nucleus. If these forces are destroyed, there would be no atom. Every organism should produce energy to live. In living beings, such energy is produced from oxidation reactions. Living beings are capable of conducting oxidation reactions by inhaling oxygen. After death, there is no possibility of oxygen intake by any organism.

    As the massive celestial body forces encircle the self-gravitational force, the nose only forms at null points for the intake of oxygen.

    If there is no nuclear force, there is no atom. In the same way, there is no existence of an organism if the nose-like structure is closed. Death will occur in a short time if there is no oxygen coming through the nose-like structure.

    Nuclear forces cannot exert their influence over longer distances. In the same way, living beings can only take in air from the nearby environment.

    We have already learnt that fishes are formed from the planets of the outer orbits in the ultra-cooled massive star body family. In the case of fishes, there is no nose, so they cannot take in oxygen from the air directly. They can only take in dissolved oxygen from the water through gills. This may be due to weak gravitational forces that are exerted by the planets of the outer orbits of the ultra-cooled massive star body family. So these fishes have stronger nuclear forces, which thus form gills behind the nose region.

    Snakes and reptiles are controlled from the far-off meteors in the ultra-cooled massive star body family. These meteors produce weak gravitational forces, and their influence on snakes and reptiles is less, thus leaving them limbless. According to this theory, snakes should also have no nose and they should be dwelling in water. Even though this is true for most reptiles, it is not true for snakes. This has to be explained in detail.

    The amphibians that are formed from the far-off asteroid belt can live both in water and on Earth. They can take in oxygen by both methods. This phenomenon is explained in detail in part II chapter 11, by the non-fixed orbits of asteroids and the fixed nature of genes in the lifetime by the biotron.

4. **Weak Nuclear Forces:** Due to weak nuclear forces, the atom will be unstable and show radioactivity. Only elements that have heavy molecular weight and artificial isotopes with unstable nuclei will show radioactivity due to weak nuclear forces, but not all elements.

In the case of organisms, only ears are formed from these weak nuclear forces. All elements do not show weak nuclear forces. In the same way, most animals have ears but some do not.

So far, we have learnt about the four independent forces. However, scientists tried to prove them as different parts of one primary force. Maxwell unified electrical energy with magnetic forces and called it electromagnetic force. Similarly, Professor Abdul Salam unified nuclear forces with electromagnetic forces and proposed the Law of Invariations, by proving that they are two facets of a single force.

If we remove the eyes and ears from any animal, it lives almost separate from surrounding organisms.

Many scientists tried to unify these four different forces as different facets of a single primary force. In a similar way, the nose, eyes, mouth, and ears originate from a single duct-like structure and later specialized into different organs.

From the information we have learnt so far, we will get the following doubts:

1. Matter is influenced by these four different forces in animals and produces the eyes, ears, nose, and mouth. If so, plants are also produced from matter. Are there any eyes, mouth, nose, and ears in plants?

2. In the case of protista, fungi, and bacteria, there is also no head, eyes, nose, mouth, and ears, even though they are made of matter. How can be this explained through biotron concept?

Explanation

The type of biofield formed for animals is shown in a diagram earlier in this part. Animals are controlled by planets and satellites, which are under the massive body fields. That is why these four forces are at null points, which are encircled by the influence of massive bodies magnetic forced lines the null point region is formed as a head with four important organs: eyes, mouth, nose, and ears.

REASONS FOR NOT HAVING PARTS SUCH AS EYES, NOSE AND MOUTH IN PLANTS

In an ultra-cooled massive star body family, the massive body is at the centre and is independent. It contains a huge amount of mass. The mass will be about 99.8% of the total of ultra-cooled massive star body family's mass, and rest of the planets, satellites, and other planetary particles occupy only 0.2%.

By studying the ultra-cooled massive star body fields, we can predict the nature of the plants field. In order to know this, we need further information about the following aspects.

Generally stable charged particles in nature are of two types: protons and electrons.

Field of a Proton

When we observe the field of a proton, we find that the forces radiate to the outer side, or periphery, from the centre, or nucleus. Here, all force lines are independent.

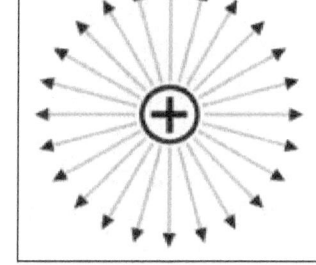

Field of an Electron

When we observe the field of an electron, we can find that the different forces are at the centre, from different directions of the outer side or periphery. All force lines are independent here, as well.

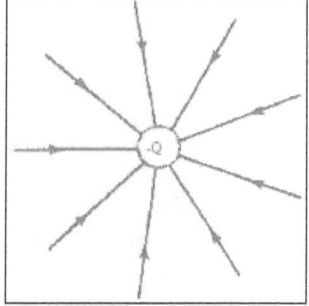

Field of a Proton-Proton Pair

Field of a Proton-Electron Pair

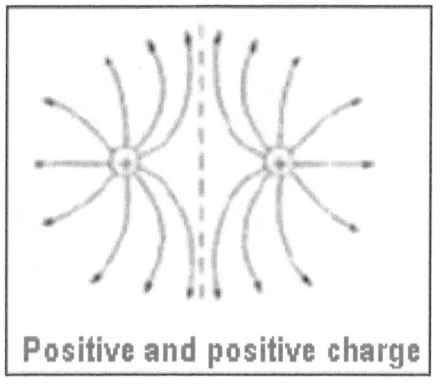

Positive and positive charge

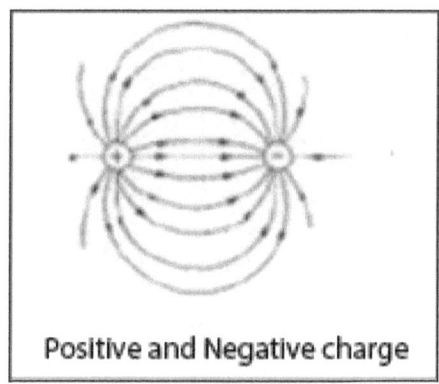

Positive and Negative charge

From the above four types of fields, we can deduce the following information. When protons and electrons are available independently, their fields are not influenced. But if electrons come nearer to protons, one influences the other and there will be change in their fields.

In an ultra-cooled massive star body family, the massive body is at the centre and it is independent and controls the formation of plants. It contains a huge amount of mass, so the massive body's field forces are strong, independent, and resembles the fields of plants.

In the case of plants' biofield, the force lines can be compared to force lines surrounding the proton. These forces radiate to the outer side or periphery from the centre or nucleus. Here, all force lines are independent. The majority of plants, except for xerophytes, herbs, and climbers, can be compared to this.

Unlike the biofield of animals, there are no null points, and field forces are open-ended. So plants need not respond to the four types of force. And every force line is independent. Because there is no null point, so the head was not formed in plants. As there are no four different field forces, there are no eyes, mouth, nose, or ears in plants.

THE BIO FIELD STRUCTURE OF A PLANT

In the proton field, the force lines are independent, produced from the centre, and move divergently. In the same way, branches and leaves are produced from the stem in all directions. Each branch is independent. If you remove one branch from a tree, it doesn't have any effect on the plant.

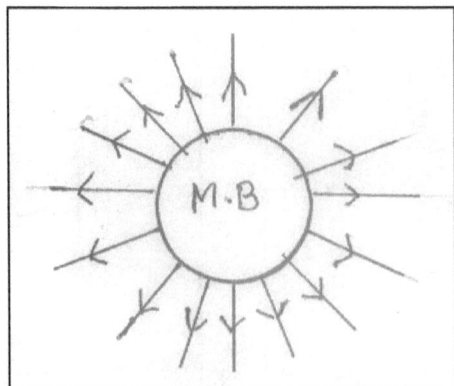

In plants, the majority of biotrons are present in the stem. Due to divergence, plants produce stems and leaves in all directions. If you remove one branch from a tree, it will regenerate in a short span of time. There is no destruction of the plant.

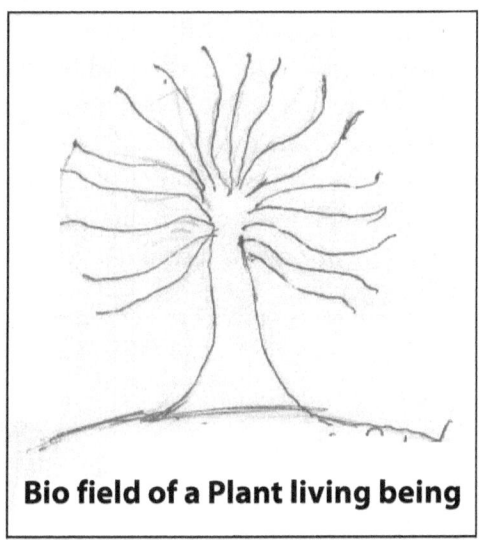

Bio field of a Plant living being

In the case of the biofield of animals, these four forces are encircled by the influence of the forces of the massive body. So if any reaction occurs in animals, it leads to the reaction of the entire animal due to the bipolar nature of the biofield. If you remove legs or forelimbs, the majority of animals will die in a short period. However, in the case of plants, two zones are present. They are the root zone and the stem zone. The reason for such zonation is given below.

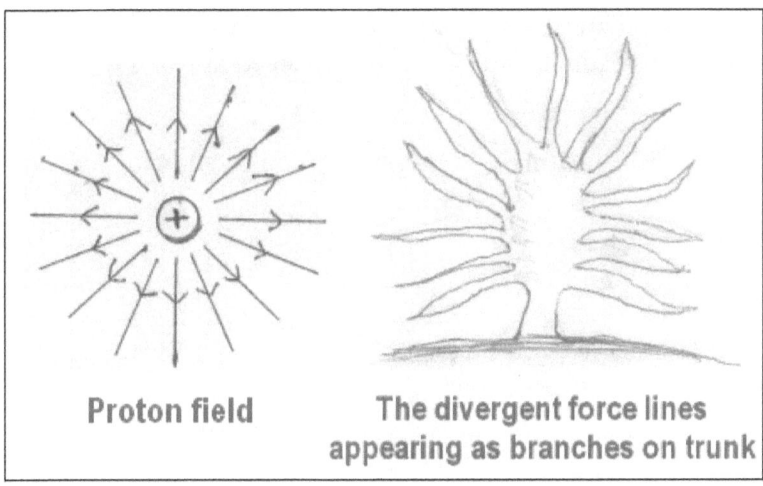

Proton field　　　　The divergent force lines appearing as branches on trunk

Positively charged protons are not static, as they rotate around themselves. Due to such self-rotation, bipolarity is formed in the positively charged proton. For example, the photon of the quantum of the energy wave also has polarity. The spin value is ½ in the clockwise direction. So in the same way, biotrons must also have self-rotation and bipolarity. One plant cell is made up of one biotron, which is equivalent to one genome. A plant is made of many cells. Due to the self-rotation of all biotrons, plant bipolarity will be formed, thus leading to the formation of the root zone in plants. This root zone is similar to branches and leaves by having a main root and side roots. This root system is also divergent, just like branches and leaves.

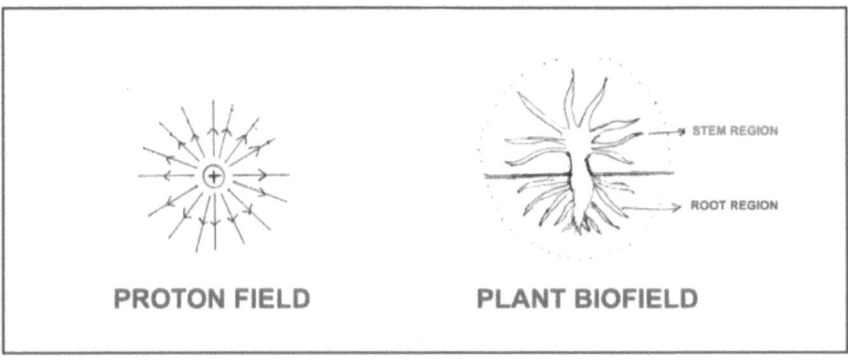

Apart from this, even a massive star body is also under the influence of distant galaxy center. This is also one of the reasons for the bipolarity of plants and the root zone formation in plants.

In the case of monocot plants, there is no main root system. The reason for such subtle variations in plants is explained in part III. The biofield in the case of protista, fungi, and bacteria is explained in part IV.

CHAPTER 7
Concept of Biotron

CONCEPT OF BIOTRON

We know that life on Earth is controlled by the ultra-cooled nonluminous star body families by producing photomagnetic waves in the form of energy packets called quanta. We coined the name biotrons for these energy quanta of photomagnetic waves, or bio-waves.

According to cell biology, the basic unit of life is the cell. Many cells together form tissues, organs, organ systems, and organisms. All of these are formed from physical matter alone. Yet they need a vital life force for their existence. For example, if you buy some amount of flesh or meat from a butcher's shop, it may contain cells, nuclei, chromosomes, DNA, tissues, organs, and organ systems, but it does not show the vital life force.

Matter cannot produce life independently. In order to exhibit life activities or vital force, matter needs the help of some other force. This vital force is derived from the ultra-cooled nonluminous star body families in the form of photomagnetic waves through energy packets called quanta. The process of producing biotrons or bio-waves is not yet established through laboratory experimentation, so we do not know the charge, spin, weight, and longevity of these biotron particles. This doesn't mean that we should suspect the existence of particles such as biotrons. Here, much importance is given to finding out the reasons for the variability of life and its existence.

According to the wave propagation law, one quantum is a primary particle. So we called it a biotron. This biotron controls life activity. If so, we have to know how many biotrons are needed to control a single cell.

The cell is the basic unit of life. The nucleus is the basis for the cell. The biotron is basis for the nucleus. If you extend this theory, we have to propose a concept of one biotron-one nucleus. Let us see some cells like skeletal muscle cells, which are polynucleated or syncytial. In certain prokaryotes, such as bacteria and viruses, there is no nucleus at all. So we cannot accept the one nucleus-one biotron concept. As all cells have chromosomes or genomes, which are made up of either DNA or RNA, we can modify this concept to one genome-one biotron. Then it will be universal and applicable to all.

Cells in the haploid state will have one set of chromosomes or genome. But cells in the diploid state will have two sets of chromosomes or two genomes. So all diploid cells are under the influence of two biotrons. The majority of cells in animals are in the diploid state. Previously, we have learnt that waves produced from high cooled planet or satellite body is divided into two parts named A and B in an ultra-cooled nonluminous massive star body. Accordingly, two sets of genomes are present in animal cells. This is true in the case of all animals. So we can confidently say that one animal cell is under the influence of two biotrons.

In the case of plants, apart from monoploids and diploids, polyploids are also present. For example, in the case of monocots (grass, rice, and wheat), a single genome is present that is each cell under the influence of one biotron. In the same way, polyploids are controlled by plenty of biotrons.

What is the basis for proposing the one genome-one biotron concept? Where biotrons are located in nucleus?, needs to be explored. All of this is explained in part II chapters 8, 9, 10, and 11.

Large numbers of cells are present in organisms. There are animals with single cells to ones with millions of cells. If that is the case, how we can tell how many biotrons are present in an organism? The possible answer for this question is the numbers of cells in an organism. In an adult human body, more than 2 trillion cells are present. So we can tell that every man has more than 4 trillion biotrons. Thus, multicellular organism is a reservoir of biotrons. We have already learnt that all these biotrons are arranged in a systematic order, according to the influence of the biofield. The shape of an organism is determined based on the forces acting on the biofield. Instead of saying that the cells are organized according to the biotrons and biofield, it is better to say that cells are formed from the biotrons and are shaped according to the biofield.

Now we have to know the state of the cell before the entry of the biotron. The waves that are produced from any atom or atomic particle self-rotate. Similarly, biotrons in the bio-waves also self-rotate, either clockwise or anticlockwise, as in the case of self-rotation of photons from light waves.

The bio-waves that are produced from the ultra-cooled nonluminous celestial bodies will enter the disk and concentrate at its focal point, the Earth. That is why life is formed on the Earth.

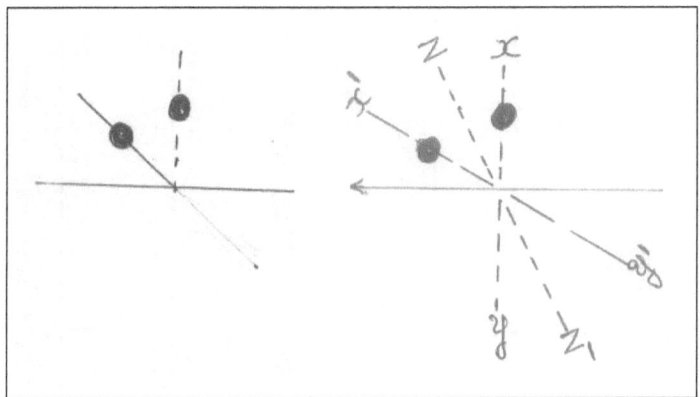

The bio-waves that are present on the Earth may be horizontal, perpendicular, or angular. The inclination of bio-waves is dependent on the celestial body that is producing these bio-waves. For example in an organism the bio-waves that are in perpendicular at the time of production will change it angle according the Biofield that are already there in organism. Here, two types of forces will work on biotrons: the force received by the biotrons from ultra-cooled celestial bodies, and the biofield force on a biotron. Due to the above forces, there will be a resultant force.

The resultant force will be between $x^1 y$ and zz^1 on the axis. What is the result of such resultant force? This has to be examined.

FORMATION OF BODY PARTS SUCH AS TAIL AND HORNS IN ANIMALS BY RESULTANT FORCE

In the biofield chapter, we have already learnt how a trunk, limbs, head, and genital organs are formed. But we did not study the formation of horns on the head and the tail at the genital region. The reason for forming horns or a tail can be attributed to the resultant force of biotrons. In any animal, we observe two horns and one tail. No animal has two tails. The main reason for this is the upward direction of the field force. As the force lines are oriented towards the South Pole due to additional forces, two horns are formed in the head region. This is similar to the formation of two hands and two legs by the influence of the massive body's force lines. At the North Pole, the force enters the hindlimbs, and the resultant force forms a tail in between the legs.

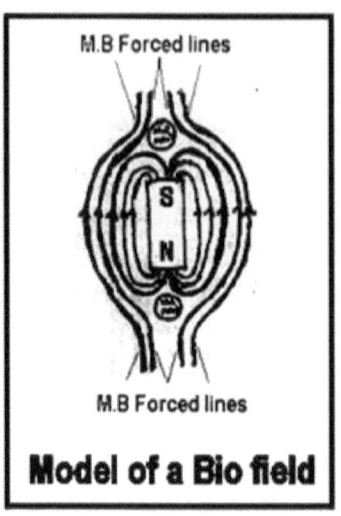

Model of a Bio field

In the animal kingdom, there are animals with horns and without horns. Goats have horns. Whether there are horns or not, the tail will be present. The rat does not have horns but has a long tail. The tail is a must. Thus, horns and tails are dependent on the amount of resultant force. Any animal can move its tail freely. Here, the force lines coming towards the North Pole are straight and without curves. There is no obstruction to these force lines. So an animal can move its tail freely. But it cannot move its horns. Here, sufficient resultant force is not there from the South Pole. In the case of man, there is no tail or horns. It means the resultant force is zero.

In the case of fishes, there are no horns, except in one or two species like whales. Apart from this, in fishes, the posterior end is split into two parts. The cooled celestial bodies that are controlling the Biofield of fishes are far away from the massive star body. So they cannot exert much influence and produce force lines in the biofield of fishes, so the resultant force is zero. That is why there are no horns or tails in fishes.

In the case of tetrapods, there are no horns at the time of birth. While these animals are growing, due to the influence of resultant forces that are newly formed in the biofield, the horns are produced. In old age, due to a reduction of the number of biotrons and concomitant reduction, the resultant force causes the horns to wither away.

When we observe the biofield, the force lines are in a long, spherical shape surrounding the median axis. In such cases, the animal should be long and spherical, and the main body axis should be in the middle. As the force lines are at the long, spherical field, the matter will also be at an axis, making it visible. In the case of large animals or chordates, a vertebral column is present on the dorsal side of the surface. In the case of tetrapods, the biofield is horizontal. The axis is inside the body. In the case of man, the axis is vertical. This is due to constant Earth gravitational force. This is explained in part II, chapter 15. In the case of man, the axis is slightly moved towards the inner side.

CHAPTER 8
Reproduction Regulated by Bio Wave Mechanism

The reason for the existence of wild nature, plants, animals, and human beings on Earth is reproduction. Reproduction is an important process. Without this, there is no existence and perpetuation of life and increase in population over a period of time.

Reproduction is essential for continuing generations and transmitting characters from parents to their offspring. From unicellular organisms like bacteria to large animals like elephants and whales of animal kingdom, and from algae to large plants in the plant kingdom, all reproduce by some method or another. Small, unicellular organisms reproduce by binary fission or multiple fission. Sporulation and budding is the process of reproduction in some multicellular organisms. Mating is an important process of reproduction in advanced and large animals.

Now, we have to explain the process of various reproductive patterns by using the concept of bio-waves in different species.

The matter that receives bio-waves gets vitality and exhibits life characteristics. Life exists at different levels, such as cells, tissues, organs, organ systems, species, genus, order, class, phylum, and kingdom. In addition to this at the individual level, life exists as monecious, dioecious, and bisexual (in plants) and male, female, and hermaphrodite (in animals). All these were explained in the chapters on astrobiology previously.

In order to explain the influence of bio-waves on reproduction, we can take one organism as an example and the concept is explained. Later, we can extrapolate this theory to other organisms.

Mechanism of Working Pattern of Bio-Waves on Reproduction

In the animal kingdom, millions of species are present. There are certain general features among them. Therefore, by studying one representative species, we can deduce and understand their general characteristics of them. Human being is also a part of animal kingdom. We know the general characteristics of human being. By taking him as an example, we will study the impact of these protomagnetic waves that are produced by the super-cooled planet on reproduction. Bio-waves produced by the super-cooled planet are influenced by the super-cooled nonluminous massive star body that controls the planet, generating the bio-waves in two types of waves, which are responsible for male and female sexual characteristics in human being. This is already explained in chapter 3 of part II. Super-cooled planets cause the variations in the sexuality of man to be limited (male or female).

In order to study reproduction, the biofields are to be examined. Matter/mass accumulates according to the field and causes the shape of the body. In the biofield, there are two null points. The biofield will be in two states: dormant and awakened.

In the dormant state at the null point of the biofield, the head region is neutralized and the cells in the head region do not receive energy. So the head swayed to the side as if it has no connection to life. In the same state at the null point of the lower side, the genital organs receive energy from the force lines directly from the field. This energy is in a reasonable quantity, and with full vigour and vitality. This can be compared to a situation where, if one pan goes up, another pan will come down to balance. All these are discussed in part II chapter 4.

In the awakened state, the force lines go upwards (↑), and the south side and head will be erect. In this state, the cells of genital organs do not receive any energy and retract with less volume.

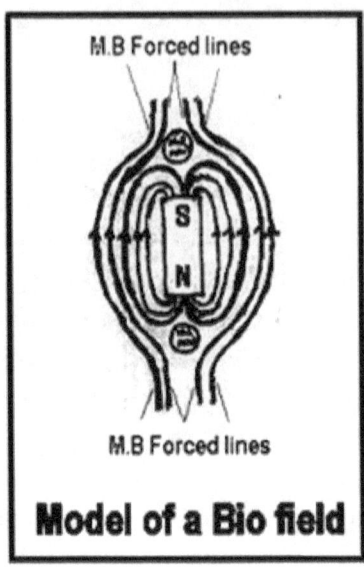

Model of a Bio field

This phenomenon can be explained with the concept of biotrons and its influence on bio-field by energy packets called quanta.

ASSIMILATION OF BIOTRON BY GENOME

In the awakened state, the force lines will work towards the south. To counteract this force, the biotrons that are controlling the cells of genital organs at the null points of the North Pole are to be moved towards the south side. We can notice the following three possibilities at this time:

1. At the north side null point, due to field forces, the biotrons in the genital part will move towards the south side bio field and to counteract this, germinal cells of the genital part have to use addition energy to hold their biotrons that is caused due to the awakened state.
2. If this is not possible, the cell should lose biotrons.
3. Otherwise, the cell should absorb or assimilate the biotrons responsible for that cell.

Let us examine the first statement, that cells in the genital part can't stop moving of biotrons due to the field force in an awakened state. So it does not occur. The second statement, about losing biotrons. If cells do not have the energy to hold biotrons, they lose their life and become dead. This is a suicidal and unwanted phenomenon.

According to the third statement, the cell should receive biotrons and absorb them. There is a greater possibility for this. There is no alternative to absorbing biotrons for the existence of cell. Now, the question arises—where does this biotron assimilate in the cell? What changes will occur in the cell due to assimilation? What are the consequences? All these are given below.

Previously, in chapter 7, we mentioned the one genome-one biotron concept. When human cells are diploid, they should contain two biotrons in each cell. According to cytology, we know that the DNA of the chromosomes of the nucleus is responsible for self-replication and protein synthesis. So the absorption of biotrons is possible only for DNA of the chromosomes in the cells.

The electrons in an atom may absorb or receive some amount of energy from photons that are present in the electromagnetic waves. But in atoms of some metals, the electrons of the outer orbital obsorb photons from light waves and emit electrons, which we call photoelectrons. This was proved by the photoelectric effect of Einstein.

Under normal circumstances, the chromosomes of the nucleus will be in the form of a chromatin network in a cell. The DNA in the form of such a chromatin network will absorb the biotron and enter the excited state by increasing the internal energy content.

The Structure of DNA

Through the use of X-ray crystallography, Wilkins and Franklin determined that DNA was double-stranded and could form a helix.

Four different de oxy nucleotides, or nucleotides, the structural units of DNA are assembled into long polymers of DNA strands, or nucleic acids. Prior to assembly, they are in the form of nucleotide triphosphates similar to ATP. Each nucleotide contains three parts: a phosphate group, the sugar deoxyribose, and one of four nitrogenous base.

The four bases of DNA, their designations and their triphosphate form are adenine A (DATP), guanine G (DGTP), thymine T (DTTP), and cytosine C (DCTP).

In 1950, Chargaff developed the principle of base pairing. He determined the relative amounts of A, T, C, and G in a variety of cells, proving that A = T and C = G, and that there are exactly as many purines (adenine and guanine) in the nucleus as there are pyrimidines (thymine and cytosine).

The Watson and Crick Model of DNA

In 1953, Watson and Crick determined the double-helix structure of DNA, including its phosphate-sugar backbone, specific (A-T, G-C) base-pairing of purines and pyrimidines, and the meaning of the intramolecular distances.

In the double helix, the two polymers run in opposite directions (5'-to-3' and 3'-to-5'). The deoxyribose sugar is the hub of this numbering system. The 3' carbon contains a hydroxyl group and the 5' carbon has a phosphate. In forming the phosphate sugar backbone of DNA, the free 3'–OH group of deoxyribose in the first nucleotide reacts with the first 5' phosphate in the second nucleotide.

Many millions of nucleotides may be present in a single DNA molecule.

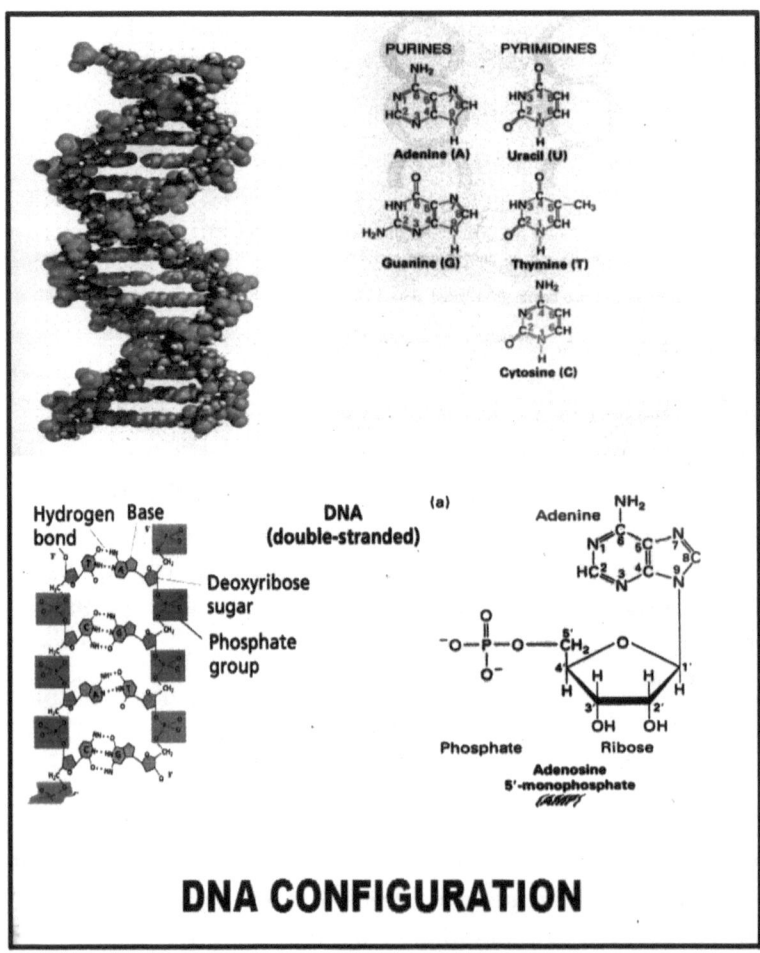

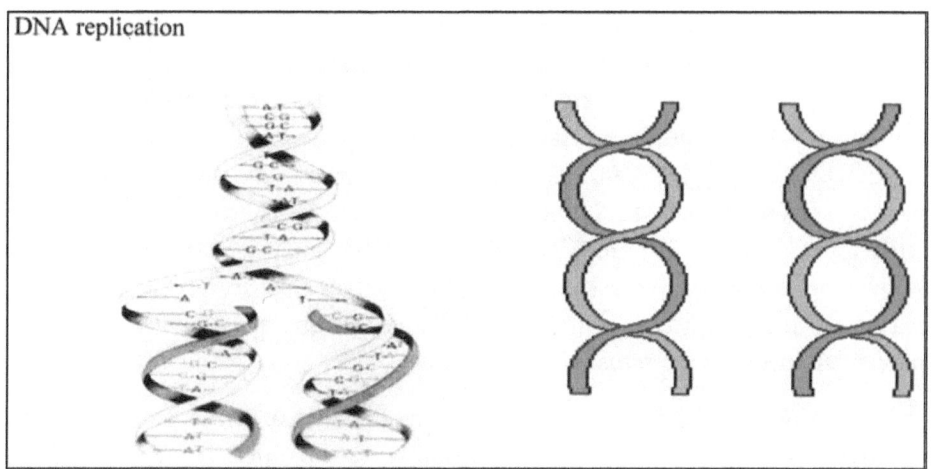

DNA replication

Replication is the preparation of DNA copies prior to reproduction of the cell or organism. Because of specific base pairing, upon separation of the DNA double helix, each strand can reproduce, serving as a template for the other. This process is called semi-conservative replication.

Absorption of the Biotron By DNA:

SELF-REPLICATION IN CHROMOSOMES BY BIOTRON POWER

Due to absorption of biotron by DNA, (due to awakened state of organism as previously said there is no option for the cell except absorption) the additional energy that is received from biotron will flow along the chains of DNA and reach the steps of the DNA ladder, the nitrogenous bases. There will be breakdowns of the hydrogen bonds, which are present in between nitrogenous bases (A = T, C = G). This breakdown of hydrogen bonds will occur until the excitation reduces to a normal level. These broken-ended nitrogenous bases are attached to the phosphate backbone on one side. In the excited state, the four nitrogenous bases that are essential for biosynthesis of DNA, along with certain enzymes, will attract the broken chain of DNA, as it has extra energy from absorbing the biotron and forms a complementary stand. Thus, the split single DNA chain will become double and form a set of two chromosomes.

At the same time, due to the absorption of biotrons by DNA, DNA is excited in the biofield, resonating with the bio-waves. There are different types of bio-waves in the vicinity of exited DNA that produce different types of biotrons. The biotron that the DNA is responding to and resonating with is considered to be the controlling biotron linked to that DNA. There are numerous ultra-cooled celestial bodies in the universe. They all produce biotrons. But one particular celestial body can excite and resonate with one particular type of DNA. We can consider such DNA to be at resonance with such biotrons or bio-waves produced by such ultra-cooled celestial body. This means only the frequency of bio-waves that are produced by one particular ultra-cooled celestial body can excite one particular DNA in a particular species or organism.

But now, the excited DNA is at the null point of the organism. Here, the field force is neutral. Apart from that, it is separated from the biofield after absorbing biotron. If an independent cell wants to receive a biotron, it can be a biotron from the relevant biofield. The reason for this is that if it receives biotrons from bio-waves, it will be according to the direction (↑) of the force lines, and so it will be attracted. It means it should receive a biotron which will neither attract nor repel the biotron. So in the biofield at the lower null point, the biotrons opposite to two poles of the field forces are to be received. Such biotrons are plenty in the vicinity, because they originate from many ultra-cooled celestial bodies. But the DNA of the cell is not formed to respond to all such stimuli. So the cell must receive only specific bio-waves, which have opposite poles from only a particular ultra-cooled astral body.

MEIOSIS IN GERMINAL CELLS DUE TO NULL POINT AREA OF BIOFIELD

Such biotrons can be produced by the waves of an ultra-cooled celestial body at an 89° to 90° inclination. For the pole movement, the cell, wave, or matter should rotate 180°. Only then will the poles change. Here, we are discussing bio-waves with 90° angles. So there is no change in the poles. But when we observe the bio-field, force lines are perpendicular to the biofield at the null point. There is already an inclination of bio-waves with 90°. So the total angle will become 180°, which is sufficient to change poles.

So the DNA will respond to biotrons produced by bio-waves with a 90° angle. Here, you have to notice that the biotron is responsible for the cell from the biofield. This shows the organism is in growth state and receiving veridical bio-waves through the disk.

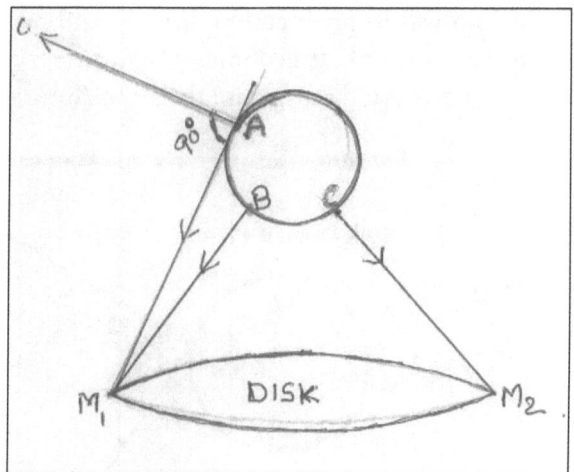

The waves required for organism is receiving from the area of B-C. So the DNA of the present organism cells is in resonance with that waves which are from the B-C region. Now it has to receive biotrons from bio-waves from region A with an 89°–90° inclination. The following questions arise after DNA absorbing biotrons from the bio-waves which are transmitted by the ultra-cooled celestial body.

1. Why should DNA respond to these biotrons that are produced by bio-waves?

Biotrons are the quanta of energy. After absorbing this energy, DNA will enter the excited state. The excited DNA will make the DNA resonate accordingly. Concomitant changes are explained in further pages.

RECOMBINATION OF DNA IN CHROMOSOME INCONSONANCE

Here we have to notice one important point. There is a change in the place and direction of propagation of waves. So accordingly, DNA has to be reorganised. There are two sets of genomes. Each set will respond to the biotrons from the bio-waves of the 89°–90° angle. Among these, the paternal genome (Y genome) will respond to the area of the male part in the celestial body which is having 89°–90° inclination. In the same way maternal genome (x genome) will respond to the area of the female part in the celestial body which are having 89°–90° inclination. It means in the two sets of genomes the DNA will respond and resonate to the two different areas of the planet with change in the bio-waves propagation to 90° and with little or no changes in the energy of biotrons. If genomes want to respond according to the different biotrons, they should reorganise their DNA.

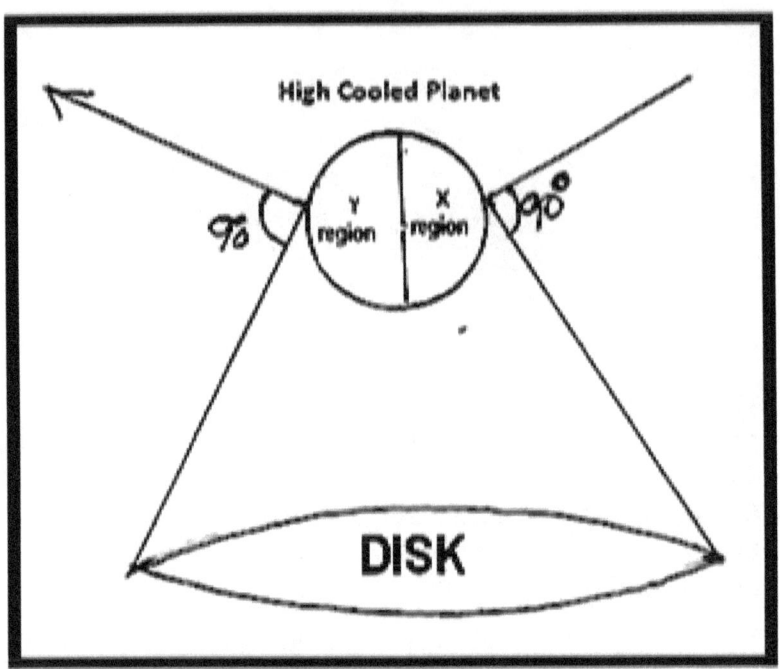

Accordingly, DNA reorganisation starts in the genome. The chromosomal part of the genome which is in resonance with the bio-waves will influence the homologous part of that chromosome in the second set of genome and reaches to its proximity by attraction. Thus, chromosomal pairing will occur. These paired chromosomes are called bivalent chromosomes in cytology. The part of the chromosomes that are in resonance with bio-waves by receiving biotrons will get responses and exchange their chromosomal bits (DNA).

CONJUGATION CONDUCTED BY BIOFIELDS

Genes are the functional and structural units of chromosomes. Genes are made up of DNA. These genes are of two types: euchromatic and heterochromatic. At times, the heterochromatic genes may change into euchromatic genes due to a change in the direction and place of bio-waves. So the DNA of the resonating chromatid in the bivalent chromosome can influence the homologous bit of chromosome or chromatid and goes near to it. So pairing occurs with homologous chromosomes. Due to this, synostomal complexes (here I used meiosis cell division terminology as it is same in this process) are formed. Based on the resonating nature of DNA, the genetic material will be modified. Chiasmata will be formed due to the proximity of the chromatid, caused due to synostomal complex formation. The state of resonance increases due to additional DNA that is accumulated on top of the previous resonance. Crossing over occurs in the parts of genes where there is more resonance. After the exchange, the bivalent chromosomes will resonate more. But here, two different sets of genomes are responding to the bio-waves produced by the two different regions of the ultra-cooled celestial body.

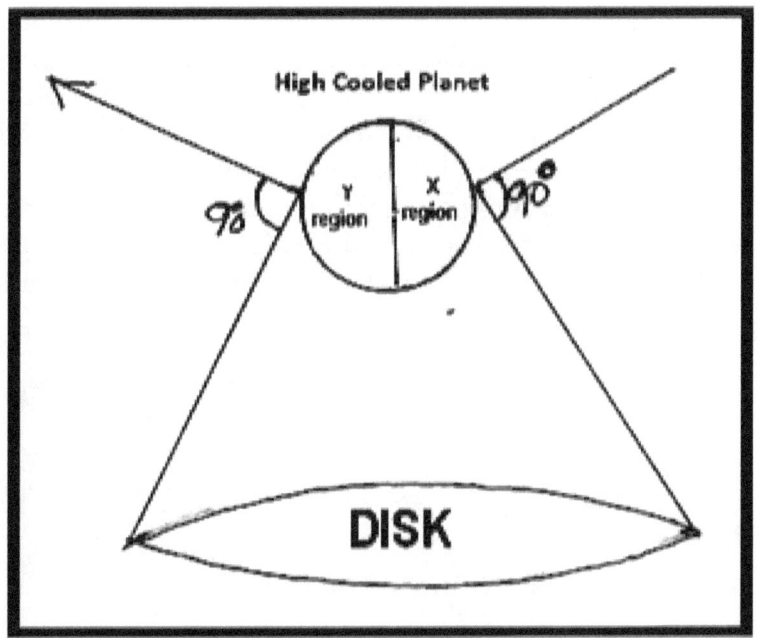

After exchange due to an increase in resonance, they try to link. The genomes start separating. The rest of the genomic region starts disjoining from the single genome. That is why the single cell will become a double cell by cytokinesis with the haploid number of chromosomes.

We have already learnt that due to the absorption of biotron, the DNA gets excited and hydrogen bonds between nitrogenous bases will be broken down, leading to the formation of a complementary strand and two chromatids in a cell. In a haploid cell, two chromatids will be present in chromosomes, resonating to the biotrons produced from the surrounding bio-waves with an 89°–90° angle.

By the influence of surrounding bio-waves with an 89°–90° angle, the biotrons that are produced will reorganise the DNA of the chromatids into two genomes due to crossing over and recombination, even though they are haploid. According to the surrounding bio-waves with 89°–90° angle, the chromatids of one side are bound. In a similar way, the rest of the chromatids will try to split at the centromere region.

With this, one side cluster of chromatids (genome) will resonate after receiving the energy quanta from biotron and get excited, giving energy to the surrounding bio-waves. The bio-waves that receive such energy will increase their momentum and angular velocity, and that changes their nature from the wave form to the particle form and they reach the genome as biotron. Further details are given in later pages. The chromatids that lost the excitation and are in the ground state will be in the form of a chromatin network. Now, the two chromatids receive biotrons in a systematic manner. Chromosomes will be in two groups, thus forming two cells. It means each haploid cell will become two haploid cells. In a similar way, one diploid cell at the null point will produce four haploid cells. When you study these cells, the following information can be obtained:

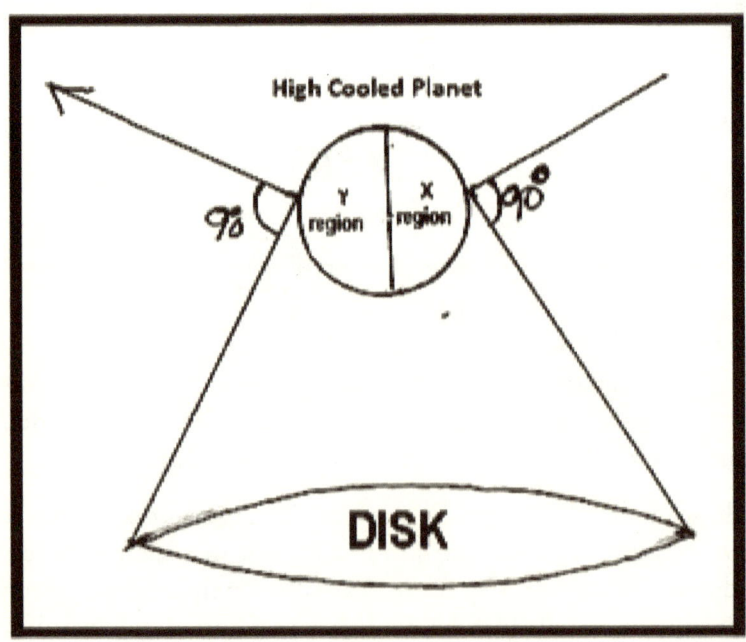

S. No.	4 Cells Formed at Null Point after Biotron Absorption	Cells in the Biofield of an Organism
1.	All four cells produced from a single cell are haploid.	The cells in the organism from where they are produced are diploid.
2.	Biotrons produced from the bio-waves with 89°–90° angle from 'A and B' regions is the basis for cells.	The perpendicular waves that are produced from the B-C region of the ultra-cooled celestial body are basis for cells of the grown individual.
3.	Due to recombination, the four cells produced are unique and independent.	All cells of an organism are under the influence of the biofield. So the genetic make-up of cells are one and the same. In terms of the cell, they are dependent. However, in terms of the organism, they are independent to some extent.
4.	Due to recombination the nucleotide sequence or genetic code is not the same in all cells.	Nucleotide sequence or genetic code is the same in all cells.
5.	Due to planetary rotation the slope decreases and the chance for receiving biotrons by the four cells increases. So there is scope for independent growth.	The organism is dependent on the longitudinal rays. So there is no change in the intensity of biotrons produced from the bio-waves. So there is no scope for development and growth.

The four haploid cells in the genital part of the null point are transformed, and later they will have independent genetic compositions. So there is variability in genetic structure from parent cells, and this is considered FOREIGN BODIES of that organism.

We have already learnt that due to planetary rotation, the slope decreases and the chance for receiving biotrons by the four cells increases. So there is a scope for independent growth by attracting matter. This matter may be from inside or outside of the body. But these four cells are present at the genital region of the null point, so there is no possibility for it coming outside. So matter from the organism slowly starts accumulating at these four cells according to the energy they receive. If this happens, it eventually leads to the destruction of the organism. To avoid such a situation, the organism has to send out these four haploid cells/foreign bodies/sperm/ova. The organism tries to push them out of the body. But that is not possible because they are in the null point region in the biofield. There organisms' biofield force lines cannot work.

An organism can send even a litre of urine or one kg of faecal matter very easily by producing muscular contractions at will. But it cannot send out a few drops of sperm. A similar situation is present in the above condition.

However, the foreign bodies have to be sent out. The exact mechanism of sending out these foreign bodies is given below. In order to understand this concept, we have to study the biofield once again.

An organism cannot expel foreign bodies through muscular force at will because they are independent. That is why they are expelled by using field forces. For this to happen, this field force should move towards genital organs. As the field force lines are upward ($\uparrow$), this is not possible. Despite that, the organism tries to expel these foreign bodies by using the internal field force from the trunk region. Due to this at the S region of the trunk and at the N region, force lines start exerting force. Due to exertion of these forces at S and N regions, additional biotrons will be added. So matter accumulates and leads to tissue growth and development. This will cause breast development in females and chest development in males. At the lower side, genital organs are developed to expel foreign bodies. As the foreign bodies are independent, they cannot be expelled very easily. Matter increases around foreign bodies due to an increase in energy, and this situation is life-threatening. To avoid this, the organism has to increase its biofield. Only then can it expel the foreign bodies easily. There is no possibility of an increase in biotrons, as the numbers of biotrons are maximum at that time due to longitudinal rays being received from the ultra-cooled celestial body (details given in part II chapter 9). So there is no possibility for a change in the biofield, and no scope for expelling foreign bodies.

The same situation can easily be tackled in the non-living system. For example, in order to increase the field force of one bar magnet, we can add another bar magnet at the opposite pole to increase the field force.

Similarly, one organism is to be in the proximity of another organism. There is no use of nearing male with male, because like poles repel and unlike poles attract. So male and female should come nearer. As we have already learnt, the ultra-cooled planetary body produces two types of bio-waves in every ultra-cooled planet. They are male and female sex-causing rays. Among them, female-forming bio-waves are negatively influenced by the biotrons from massive bodies, and pseudo poles are formed additionally. So due to the union of male and female, the biofield will be enhanced. The foreign cells can be sent out very easily.

So far, we have explained the existing situation in males. A similar situation may occur in the case of females as well. Here, foreign bodies will be in the form of ova. The same situation will occur in the female. In the case of males, from one sperm mother cell there are four sperm cells formed. But in the case of females, from one ovum mother cell one ovum will be formed and the other three will die. The reason for this is explained later. In both males and females, a similar situation takes place with little variation.

In order to increase the biofield, both male and female should unite. How does an organism know this? Who will give the expert advice? Such doubts will occur naturally to anybody.

In reality, the internal conflicts that occur in males or females can be observed by changes in the secondary sexual organs. We can find the change in biofield by touch, proximity, vision of a male with a female. In order to realize such changes, one sex reaches out to another sex. After mating, the biofield force increases, and foreign bodies will be expelled.

In the case of males, the field force is more due to the positive result of bio-waves produced from the ultra-cooled massive star body. But in the case of females, the field force is less due to the negative result of bio-waves produced from the ultra-cooled massive star body. But at the time of mating, male and female fields are united, and due to resultant enhanced field forces, the male will expel the sperm/foreign bodies at the field of neutral point of the female and will withdraw from the common biofield of the male and female (intercourse).

But in the case of a female, the situation is different. The field force is less. By utilizing the field force of males after expelling the foreign bodies from mating with the male, it should expel the ovum. The male withdraws immediately after mating but, due to the slower action of the female, she cannot expel the ovum quickly. So the single way of sending the ovum to the outside is closed. But the additional field force received by the ovum at the time of mating causes momentum in the ovum.

So far, we have learnt about the diploid cells and haploid cells (sperm and ovum) and their different stages of production, development, and removal. But when you interpret this information in terms of biology, we have to answer some more questions that arise from it.

Q1. In this chapter, it was stated that biotrons are absorbed by the DNA and got excited? What is the basis for this? Where is the biotron in the cell? How do we identify it? What is the situation if it is present or not?

NUCLEOLUS OR PLASMOSOME, WHICH HOLDS THE BIOTRON?

Ans. The biotron is an energy packet or a minute particle, like a photon or phonon. So far, there is no instrument on Earth that shows the photon. But there are plenty of instruments that show the existence of the photon. In the same way, biotrons cannot be seen, but their existence can be easily recognized.

The place of biotron is the nucleus. Biotron is absorbed by DNA. DNA is present in the chromosomes of the nucleus. There are two biotrons in the diploid cell, and each biotron controls a single genome. DNA gets excited after receiving the biotron and breaks down the hydrogen bonds in between the nitrogenous bases, causing repulsion. After separation of two strands, the energy received at the phosphoric acid, deoxyribose sugars, and nitrogenous bases will be useful for making complementary strand. Thus, self-replication occurs.

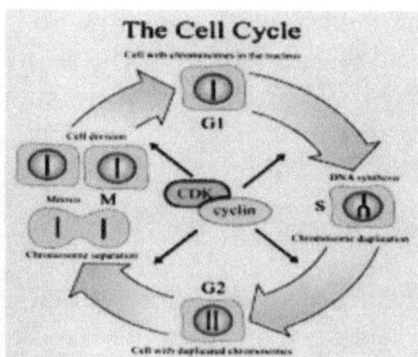

All these processes will occur at the time of cell division. At this time, all parts of the nucleus are clearly visible. After cell division, they disappear. The location of the biotron in the nucleus is the nucleolus (another name is plasmosome), or the plasomosome. At the time of cell division, due to absorption of biotron by the genome, the nucleolus will disappear and will only appear after cell division due to the reestablishment of the biotron in the nucleus. The reason for this is given below.

According to modern cell biology, the self-replication occurs in the S phase of the inter phase of cell cycle. The biotron might have been absorbed by the genome only at this stage. That is why, after receiving energy from biotron, the DNA gets excited and splits, leading to self-replication. According to this, the nucleolus should disappear only at the S phase of the cell cycle.

But according to the knowledge of modern cell biology, the nucleolus does not disappear at the S phase. Instead, it disappears at the last moments of the prophase and appears at the telophase. This can be explained with the following reason.

The chemical composition of the nucleolus is obtained by doing biochemical tests. The nucleolus contains RNA, histones, phosphor-proteins, enzymes, and salts like calcium, magnesium, and zinc. Ribosomes of 50–100 A° units and filamentous, wool spindle-like nucleonema are also present in the nucleolus. The biotron is a micro-energy particle. After a biotron is absorbed by a genome, the biotron may disappear, but not all the matter in the nucleolus.

For example, if an individual dies, his body will not disappear immediately. Instead, it slowly degrades and the matter merges with nature. In spite of the biotron disappearing at the S phase, the particulate matter of the nucleolus gradually dissolves into nuclear sap until the ending of the prophase in the cell cycle.

After DNA absorbs biotron in the S phase, it will excite and form the complementary strand of DNA. There will still be energy waves (biotrons produced through the disk) in the surroundings of the DNA that lead to the excitation of DNA. Due to this, the second excitation of DNA should cause it to split again. But this does not occur in the cell. So the DNA should withstand the energy that was additionally received from the biotrons at resonance. There should be a method to circumvent this problem.

For example, if you connect a wire to an electrical circuit, it conducts electricity from one end to the other end. Instead of that, if the same wire is coiled like a spiral and connects to an electrical circuit, it conducts electricity not only along the wire but also in other directions. From this, we can conclude that a coil or spiral wire can withhold a certain amount of energy.

SPIRALISATION AND CONDENSATION OF CHROMOSOMES DUE TO RESONANCE

In the same way, the additional energy received by DNA from bio-waves due to resonance will also be used for spiralisation of the DNA, and will prevent the damage of the DNA due to receiving additional energy by resonance.

From the time of the formation of the chromatids in the S phase of the cell cycle, condensation occurs in the chromatin network. It indicates exiting by receiving energy due to the resonance of DNA by the surrounding bio-waves. Due to excitation of the DNA, it gets condensed and spiralisation starts. This resonance is not transient but continues until the gap-2 phase (G2 phase) of the cell cycle.

Spiralisation of DNA occurs in two types: the Paranemic type and the Plectonemic type. The rate of spiralisation is proportional to the amount necessary to establish a biotron for a genome. DNA spiralisation prevents the splitting of DNA. If the resonance is higher in DNA, there will be a special type of spiralisation with additional loops, which can be seen in the lamp brush chromosomes of chironamous larva. Additional loops will be formed in the spiral as shown in the diagram.

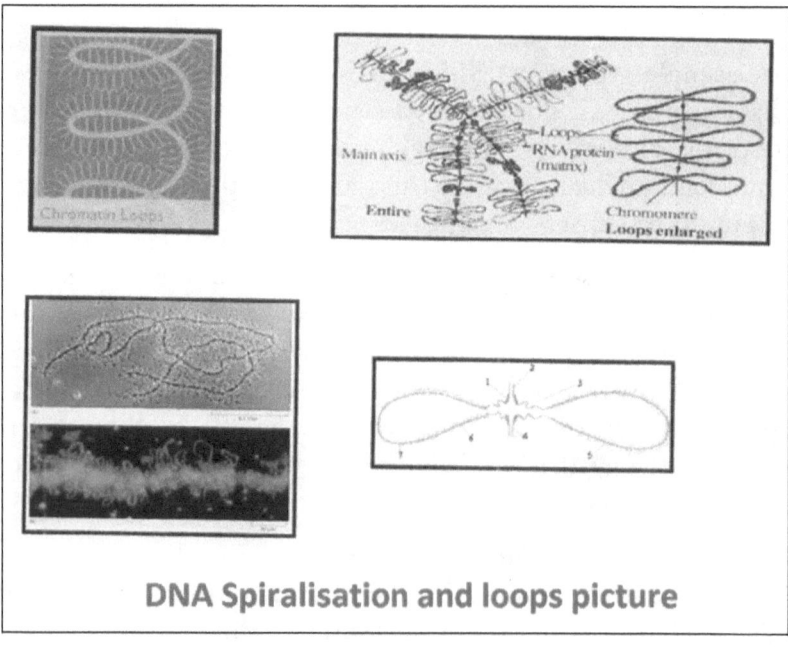

DNA Spiralisation and loops picture

Particular DNA can respond and resonate to only a specific frequency of bio-wave resonance, which is specific to the frequency of the bio-waves of a particular ultra-cooled celestial body. Chromosomes are shortened due to condensation caused by the resonance of DNA. At this stage, the energy received by the DNA will merge with the genome due to attraction. This can be called the beginning of cell division. Now, the nucleolus of the cell, which will lose its biotron and physical matter, will be distorted and dissolve in the nuclear sap due to the movement of chromatids.

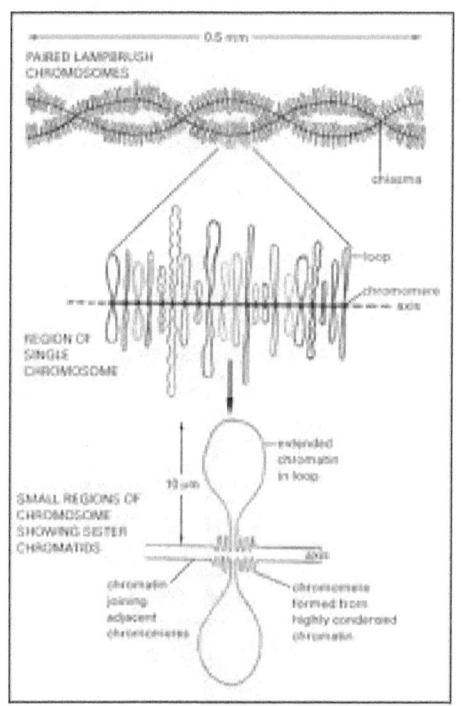
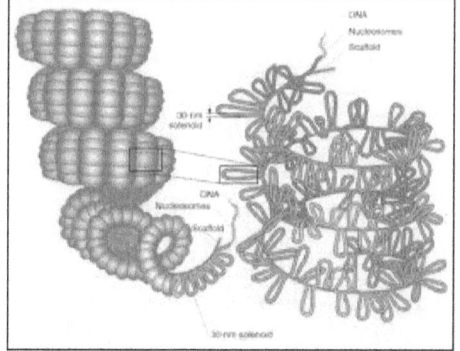

After the excitation of DNA due to the absorption of the biotron from the bio-waves in the surroundings to hold a biotron, the chromosome starts splitting, and thus chromatids will unite. Here, the Anaphase of the cell cycle starts. Then chromatids will separate from their sister chromatids and move towards biotrons. Thus, the released daughter chromosomes will resonate in accordance with the bio-waves.

FORMATION OF BIOTRON DUE TO CREDITED POWER BY SPIRALISATION OF CHROMOSOMES

The condensed and spiralised daughter chromosomes are exposed to any one of these three situations:

1. Receiving resonating biotron and getting excited.

2. Resonating to the surrounding bio-waves without any absorption of biotrons.

3. Capturing the biotrons that are passing on it and holding them as a support.

Now let's study what will happen when these situation occur:

1. The daughter chromosomes should absorb the biotron that are passing on them, but it doesn't happen. This is because it is already excited, and if it absorbs an additional biotron, the internal energy level will increase more spontaneously and it will immediately split the double helix strand. But such a situation does not occur in the cell.

2. If it is resonating to the surrounding bio-waves without any absorption of biotron, it should always be in a condensed state. Such a situation is not observed.

3. Capturing the biotron that are passing on the chromatid and holding them as a support is possible. Then, the daughter chromosomes can lose their excitation by transferring the energy to the biotron and chromatids into the chromatin network by de-condensation.

A biotron will be captured by the chromosomes, which will transfer their excitation to the biotron, before the chromosomes are converted into the chromatin network. This is exactly what is happening in the cell. In confirmation of this, at the place of biotron, the nucleolus appears after chromosomal de-condensation at the telophase. A diploid cell needs two nucleoli and a triploid cell needs three nucleoli to capture the biotrons and transfer them to the nucleolus. This is in conformation with studies of cell biology.

So far, we have explained life at the atomic, molecular, and cellular levels in terms of physical forces and physics laws, and various factors influencing it. But we do not know how a daughter chromosome will excite and capture a biotron that is in the wave form. This is to be explained in terms of physical laws.

Explanation

The entire knowledge of physics will explain about two important aspects: matter or mass, and radiation or waves. These two are the basis for energy. The whole universe is filled with matter and its radiation. Even though the characteristics of matter and waves are different, matter will not only show the properties of matter but also the properties of waves.

DIBRAGLEE DOCTRINE – INTERATIONS OF MATTER AND WAVES – BIRTH OF BIOTRON

For example, we know that electrons, protons, and neutrons are made up of mass. But according to the De Broglie experiment, they also show wave properties. The same mass at lower velocities shows wave properties and at high velocities behaves like particles. For example, we know that a neutron is a particle. But its average velocity is less than 2.2×10^5 cm/sec, it shows its wave property.

According to physics, we know that waves also show a particle nature. If waves interact with waves, they show wave nature, and when waves interact with matter, they take on the particle nature. If a wave interacts with another wave, it shows the property of interference. If a wave interacts with matter, it shows particle nature. For example, if light waves fall on metallic atoms, waves convert their wave nature in to particle nature named Photons by these photons pressure photo electrons are produced. This establishes the particle nature of waves.

In the case of the cell, the coiled and excited genome (set of chromosomes) is nothing but matter. The surrounding protomagnetic waves or bio-waves are in wave nature. Due to interaction of the bio-waves with the matter of the genome, the bio-waves lose the properties of waves and exhibit the properties of particles. This is why the energy packets of the bio-waves are called biotrons.

Thus, the conversion of the wave form of biotron to the particle form of biotron is possible by a change in the velocity of bio-waves. This is only possible due to a change in the energy content. If the energy content increases, the velocity increases. Due to high velocity, the biotron changes its nature from a wave to particle.

So the already spiralled and excited genome, which is in resonance, will receive energy from the surrounding energy packets. Then, the biowave's velocity increases and converts it into a biotron particle that adheres to the genome. By losing energy, the DNA will be de-condensed and form the chromatin fibers. As a mark of this, the nucleolus will be formed at the place of the genome. One chromosome of the genome is called the nucleolar organizer chromosome in cell biology. So the nucleolus can be considered the place of the biotron.

If you give energy to a moving particle, its velocity will increase. Due to high velocity, the energy packet or biotron will exit. We don't know the reason why this biotron will adhere to the genome. The answer for this question is given below.

There will be the possibility of an increase in linear velocity, or momentum, or rotational velocity due to the liberation of the biotron from the highly condensed, excited chromosome with high velocity.

If you hold a tennis ball between two fingers in one hand and hit the net horizontally with the other hand, its rotational velocity increases and it revolves in the air for some time.

If you fire a metre-long Diwali cracker that is in the form of a geocycle by straightening the first 5 cm, it will move with linear speed until it exhausts that 5 cm length, and while burning the spiral part it will rotate with rotational velocity and it appear in the same place even though it is rotational movement.

In the same way, the biotron is bound in the genome without escaping due to its rotational velocity.

Q2. If the nucleolus is the basis for the biotron, what about prokaryotes like bacteria, which do not have any nucleus?

Ans. DNA is the basis for the nucleus. Even though prokaryotes do not have a nucleus or nucleolus, DNA is eminent. So the basis for biotrons in them is DNA. Further details are provided in part IV.

Q3. DNA gets excited after receiving the biotron and breaks down the hydrogen bonds in between nitrogenous bases and cause repulsion. After two strands separate, the energy that is received by the phosphoric acid, deoxyribose sugars, and nitrogenous bases will be used to make the complementary strand. Thus, self-replication occurs. But biotechnology and cell biology state that DNA is broken down by the enzymes, like restriction enzyme and endo-nucleases, and the complementary strands form due to the action of enzymes like ligases. Explain.

ENHANCEMENT OF THE IMPORTANCE OF NUCLEOLUS WITH IMBIBITION OF BIOTRON

Ans. Five decades ago, the nucleolus was considered to be an unimportant cell organelle. Much significance was not given to it. But now it is considered to be the most important organelle, which controls cell division as well as protein synthesis. The majority of restriction enzymes, endonucleases and ligases are produced in the nucleolus. Since the basis for the nucleolus is the biotron, we can imply that the biotron is synthesizing these enzymes. These enzymes are useful for catalysing DNA-splitting and DNA-replication. As these enzymes are synthesized in the nucleolus, we can interpret that the biotrons are responsible for the synthesis of such enzymes.

In 1971, famous scientist Dr. Daniel Nathans discovered endonucleases. So far, more than 100 endonucleases have been identified, isolated, characterized, and crystallized. But all of these are from prokaryotes. None of these are isolated from eukaryotes. The reason for that is unknown. So far, we have discussed the changes of DNA in eukaryotes but not prokaryotes. So this data cannot be extrapolated to eukaryotes. Even though the above information seems to be reasonable, the main reason is that at first, the energy receives molecules from the biotron and forms enzymes. Through enzymes, DNA genes respond, forming the template of RNA. This full information is given in part IV.

In This Chapter of Reproduction, We Have Explained the Following Concepts

1. Formation of four haploid sperm cells from a diploid cell by rays with 89°–90° angle.

2. How these four haploid cells are foreign in terms of animals.

3. The method of expelling these four foreign cells/ovum by enhancing the biofield after mating with the opposite has to be explained.

4. Males expelling sperm due to positive action of the force by the massive star body.

5. Females not expelling the ovum due to negative action of the force by the massive star body.

6. Non-expulsion of ovum in female due to delayed activity after immediate withdrawal of male from the mating or common field force.

But we have not explained the process of gestation, pregnancy, delivery, and child birth. They are given below in detail. Here, we will discuss one important question that cannot be answered by a scientist and that is the most pertinent and basic for the existence of life.

The cells that are produced haploid are diploid. That too, they are dependent on the waves produced from the B-C region through the disk, which is longitudinal. The energy content in these waves is maximum, so they get the maximum growth. There is no further scope for development for want of an enhanced biofield.

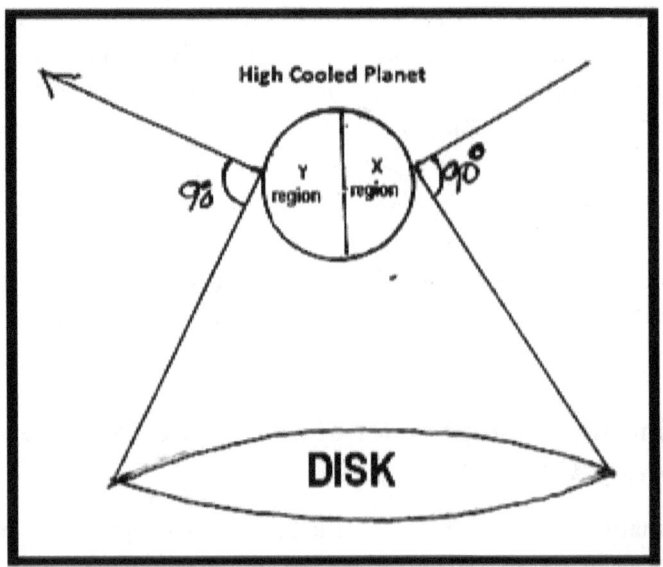

Haploid cells are formed after receiving the biotrons from rays of 89°–90° angle. Due to gradual planetary rotation, the slant of the rays decreases and the number of biotrons increases. This will give scope for transforming them into multicellular systems. They are at the null point in the biofield. If matter moves into these haploid cells, the organism cannot do anything and takes a spectator role.

In this state, accumulation of matter is possible due to enhanced energy in these haploid cells. They also have one complete set of genomes for full-fledged development. So in that field alone, they will grow as an embryo.

But in reality, they are not growing in that field. They would prefer to die rather than grow there. This is the million-dollar question facing the scientific community. The reason for such behaviour is explained below in terms of bio-waves.

DOPPLER EFFECT

Haploid, Diploid and Polyploidy by Doppler Effect

Doppler effect means the frequency of waves coming from an object moving from a point towards fixed observer will be greater than the frequency produced. In the same way, if an object is moving away from a fixed observer, its frequency is less than the frequency produced.

By applying the Doppler Effect, we can measure the distance between Earth and a star and find out whether that star is coming towards the Earth or going away from it. If it is coming towards the Earth, its frequency increases, which is called a blue shift. If it is going away from the Earth, its frequency decreases, which we call a red shift.

The same situation is happening to foreign cells like sperm or ovum. These cells are dependent on the rays that are produced from the A region with an 89°–90° angle. Gradually, the inclination of the rays decreases and the intensity and number of biotrons increases. Consequently, the frequencies of these biotrons increases and their wavelength decreases. So DNA has to be reorganised according to the changing frequency of biotrons or it must lose its resonance.

In the beginning, these foreign cells are part of the cells of an organism. They are at the null point and are later separated from the direction of the field forces. In order to resist the attraction of the biofield, they received energy from biotrons of longitudinal rays with less energy. According to the resonating frequency of slanting waves, the chromosomes are paired as bivalent and homologous synapses. Based on the resonating frequencies of DNA, there is a crossover by an exchange of bits of chromosomes or recombination of DNA. So the DNA is reorganised according to the frequency of slant rays. All this happens in an individual foreign cells' form. But within a short time, they have to face a peculiar situation.

In the reorganisation of DNA, due to Doppler effect, the planetary rotation, frequency increases in bio-waves, ultimately leading to the increase in number of biotrons or decrease of wave lenth. Again, the DNA has to reorganise in accordance with the changing frequency of the bio-waves. For reorganising DNA, there should be homologous chromosomes in the vicinity. At the diploid state, one maternal genome and one paternal genome are present. But these foreign cells are haploid. To avoid this Doppler effect, the foreign cell should change to the diploid state. The foreign cells do not utilize the chance of converting into multiple cells. This situation applies only to the foreign cells, and not to the individual. So the individual tries to expel these foreign cells out of the body.

Now we have understood the reason for the question of why these foreign cells are not growing despite having energy and matter in the vicinity. The immediate job of these foreign cells is to resist the Doppler effect by pairing with another set of homologous chromosomes. If this does not occur, there is no possibility of growing these foreign cells into multiple cells, despite having plenty of matter in the surroundings.

To avoid the Doppler effect, one sperm may unite with another sperm, thus getting the diploid state. But fertilization does not occur by the union of one sperm or foreign cell with another sperm. Both sperm cells are controlled by the A region of the planet with the production of bio-waves. As the place, phase, and direction are one and the same, the biotrons produced from them will repel and cannot be united. Even though plenty of genomes are nearby, they cannot unite. It is something like nothing amid plenty.

PHYSICAL PROPERTIES OF BIOTRON DIRECTING THE FERTILIZATION

Under these circumstances, due to the differential polarity of males and females, the biofield will be increased, and with this unified field (enhanced field), foreign cells will be expelled. Thus the sperm cells expelled will be at the doorsteps of the neutral area of the female genital organs. Unlike poles will attract. Sperms will go towards the ovum. There is no resilience or hindrance for the entry of sperm into the ovum, as this area is neutral. Thus, the sperm unites with the ovum, which has homologous DNA. Thus, a zygote will be formed.

Now, the zygote has the capacity to resist the Doppler effect. So in females, the cells start taking in matter and, after repeated cell divisions, the zygote will be converted into an embryo. Due to enhanced rotational speed, there will be an increase in the number of biotrons, which causes the repeated cell divisions and fast growth of embryo.

Here, we have to notice that with the help of the male field force, the female field force is increased and cannot expel ovum or foreign cell. Apart from this, the sperm cells are allowed into the ovum and fertilization has taken place by forming a zygote.

Mass increases in the embryo over a period of time due to the absorption of biotrons. The increase in mass is higher at the early period of pregnancy than the later period. The rate of cell division is higher in early periods of pregnancy. As the biomass increases, the embryo will occupy the neutral place (null point area) of female genital organs. If biomass increases further, it starts occupying the bio field of females just above the genital organs. With this, the female will try to expel the foetus, which enters into the periphery.

INNER FORCE LINES OF A BIO FIELD CAUSING PARTURITION

As the number of cells increasing in the foetus, the bio field of the foetus increases. We know that the inner force lines of a field are stronger than the outer ones. When the foetus tries to enter the inner force lines of the female bio field, the female will try to expel the foetus with strong force lines that are present inside. At this stage, the number of biotrons increases at both poles of the female's field. At one pole, the breasts will develop due to enhanced biotrons. At the lower pole, the enhanced biotrons are useful for expelling the foetus to the outside by using their energy. Due to this, there is no additional growth of tissue at the lower pole.

When the field force of a female increases, more than the field force of the foetus, the foetus will be expelled. This process is called delivery or parturition.

The following questions will arise after studying the above concept.

Q1. To withstand the Doppler effect, haploid cells convert into diploid cells. To withstand the Doppler effect, the unicellular zygote is converted into a multicellular embryo. To withstand the Doppler effect, DNA was reorganised in accordance with biotrons. But cell biology states that in a growing embryo, there will be mitotic cell divisions. There is no recombination in mitotic cell divisions. If so, how will the DNA be reorganised? The objective of the haploid cell is to change into the diploid cell by reorganising DNA. Instead of that, it followed mitosis. What is the reason for this? Why did the foreign haploid cell not grow into an embryo by taking on mass from the bio field where it first generated?

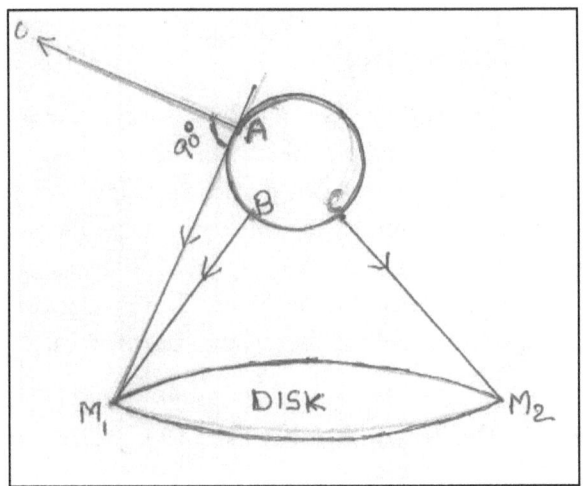

Ans. The fully grown man is under the control of biotrons produced by the vertical rays from the BC region of the planet. The four foreign cells form biotrons when the rays are at an 89°–90° angle from the A region. While going from A to B, there will be a gradual reduction in the slope, and straight-line rays will be produced. At 0 slope, there will be maximum energy and growth is maximum. In the case of man, it takes 20 years to reach from A to B. Longitudinal rays will have maximum energy and intensity. To adopt to such sudden changes, the foreign cells will try to reorganise their DNA. This is the first experience faced by the foreign cells. After this experience, there will be a change in the intensity, number, rotational velocity, and Doppler effect on the biotrons. However, these changes are small. These changes are given below in terms of mathematical calculations.

BIO-RAYS CAUSING MEIOSIS AND RELEVANT MATHEMATICAL DATA

The time required for wave to travel from 90° slope to 0° = 20 years

The angle that was reached over twenty years = 90°

The angle that was reached over one year = 90°/20

The angle that was reached over one month = 90°/(20 × 12)

The angle that was reached over a day = 90°/(20 × 12 × 30)

The angle that was reached over 6 days = (90° × 6)/(20 × 12 × 30) = 0.075°

(Because the lifespan of sperm is 4 to 6 days.)

In order to reorganise the genetic material (DNA), the DNA forms chiasma with the resonating part of the homologous chromosomes after synaptic pairing.

The number and locations of chiasma formation on a chromatid are to be determined.

In the case of man, there are 23 chromosomes and 30,000 genes. Out of these 30,000 genes, 20% are active Euchromatic genes. The remaining 80% of genes are inactive, heterochromatic genes.

The total genes present in man's genome

$$= 30,000$$

The average number of genes in the chromosome of a man's genome

$$= 30,000/23 = 1304.4$$

Total active genes present in the chromosome of a man (20%)

$$(30,000 \times 20)/(23 \times 100) = 260$$

This means that in these 260 genes of a chromosome, only transfer of chromatids will occur at the place of chiasmata in active euchromatic genes by crossing over in accordance with resonance of DNA. None have any knowledge or theory in Biology which tells about the number of chiasmata that are going to be formed in a chromatid. But famous scientist Pritchard predicted that only at a low number of places of chromatids of homologous chromosomes will crossing over occur. This is taken as basis for formation of chiasmata. By leaving the minimum and taking the maximum number that is 25% of genes (260 × 25/100 = 65) if we calculate at least 65 chiasma are to be formed in chromatid at the time of formation of a foreign cell/sperm/ovum.

So there is possibility for exchange of genetic material at 65 places (chiasmata) in a chromatid to reorganise the DNA due to the influence of biotrons producing from the bio-waves of 90° angle.

After formation of foreign cells (sperm), they will live only for 4 to 6 days. Within that, they should periodically absorb biotrons and convert into two cells. We already know that there will be a change in the intensity, number, and rotational velocity due to Doppler effect on the biotrons. This can be calculated as follows:

Change in inclination of angle over 20 years

$$= 90° - 0° = 90°$$

The number of chiasmata formed in a chromatid

$$= 65$$

The life span of foreign cell (Sperm)

$$= 6 \text{ days}$$

Change in inclination of angle over 6 days

$$= (90° \times 6)/(20 \times 12 \times 30) = 0.075$$

The number of chiasmata formed for 90° angle

$$= 65$$

The number of chiasmata to be formed for 1° angle

$$= 65/90$$

The number of chiasmata to be formed for 0.075° change

$$= (65°/90) \times 0.075 = 0.054$$

From the above statistics, it is clear that the number of chiasmata that form within 6 days in a sperm cell is 0.054.

In order to increase the volume and become double-celled, the foreign cell does not need any chiasmata or genetic reorganisation of DNA. In such cases, can the foreign cell grow as an embryo in the male itself?

Due to the change in the rotational velocity, intensity, and phase of biowaves, the genome of the foreign cells (sperm/ovum) will try to reorganise their genetic material with the help of homologous chromosomes. In such circumstances, the foreign cells will reject all other opportunities and try to convert into diploid cells. Later, by accepting matter, the cells will divide many times. In each cell division, the chiasmata to be formed is calculated as 0.054. There is a meaning if you say chiasmata is formed at one or two places. But there is no meaning if the number of chiasmata that are formed is less than one or zero. So the proposition of chiasmata at each cell division is baseless.

At the time of cell division in the zygote after taking matter, a cell will divide into two by mitotic cell division without forming any chiasmata. So in mitosis, there is no need of reorganisation of genetic material. This is explained in part II chapter 9.

Q2. The sperm/ovum mother cells are in the diploid state. They have two sets of genomes, one derived from the mother (rays coming from the half part of the planet i.e. B region back side) and the other from the father (rays coming from the half part of the planet towards i.e. A region celestial body). Out of the four foreign cells produced from a mother cell, two cells are made from the rays coming from the male region of the planet containing the male (Y) genome, and the other rays come from the back of the planetary body containing the female (X) genome. Out of such millions of foreign cells, exactly half will contain the male (Y) genome and the other half contains the female (X) genome. This means that in the vicinity of every foreign cell, there are plenty of homologous foreign cells. To avoid the influence of planetary revolution and the Doppler effect, one foreign cell containing the Y genome may unite with a foreign cell containing the X genome, and embryogenesis can be established instead of seeking for heterologous foreign cells from the female.

X-GENOME SPERMATOZOON AVOIDING FERTILIZING WITH Y-GENOME SPERMATOZOON

In reality, foreign cells of the ovum containing the X genome unite with foreign cells of males containing sperm. In sperm cells, 50% contain X genomes and 50% contain Y genomes. The reason for this is to be explained in terms of bio-waves concept.

Ans. Sperm cells may contain X (50%) or Y (50%) genomes, but all ova contain only the X genome. If the ovum (X genome) unites with a heterologous sperm (Y genome), all the zygotes formed are male (XY genomes). There is no possibility of forming females because in a cell, homologous biotrons (XX genomes) will repel and separate into two. But in nature, we find 50% males and 50% females.

Out of four foreign cells formed, two are from the maternal genome (X), which is weak in terms of energy, and two are from the paternal genome (Y), which is strong in terms of energy. These foreign cells are to be expelled either by using muscular energy or field force. Foreign cells containing the X genome have weak energy, cannot resist any field force, and are very easily expelled. Foreign cells (sperm cells) containing the Y genome have more energy, resist any field force, and cannot be easily expelled. In the case of foreign cells containing X and Y genomes, the body separately tries to expel foreign cells containing the X genome normally or by enhancing the field with the female.

The X genome contains foreign cells, when they are close to a female, they unite with the ovum and produce a female zygote with XX genomes. All produced from it are females.

The X genome containing foreign of male when near a female, will be expelled without resistance by using the self-field force. They will all die in the external environment without any adaptation.

The rest of the Y genome-containing foreign cells they are expelled out from the male when close to a female due to united field force of male and female and will be liberated near the female genital organs and unite with the ovum to produce a male zygote with XY genomes. Here, all of them are will be males. In reality, it does not happen like that. The reason for such a situation will be explained by using bio-waves. To overcome this critical situation, we have to adopt a postulation.

Postulation

At the time of sperm cell formation in males, the X maternal genome has to receive biotrons from the waves with an 89°–90° slant from the male region of the planet, instead of receiving biotrons from the waves with an 89°–90° slant from the female region at the back of the planet.

Symbol for biotron originating from male region

$$= y$$

Symbol for biotron originating from female region

$$= x$$

According to normal behaviour, the biotron will be received by the maternal X genome.

Biotron type form of the foreign cell

$$= X^x$$

(Here, X = maternal genome; x = biotron from the female region)

According to the postulation, the biotron is to be received by the paternal X genome.

Biotron type of the foreign cell

$$= X^y$$

(Here, X = maternal genome; y = biotron from the male region)

How will this new postulation overcome the problems in the case of the foreign cell? This is to be explored.

X-GENOME OVUM IS ABLE TO FUSE WITH X-GENOME SPERM CELL

According to the above postulation, all sperm cells (containing either X or Y genome) contain biotron y, as the waves come from the male region. The energy form of these foreign cells is $Y^y Y^y$. So the basis for all foreign cells in males is biotron y. Homologous cells for these are $y \times y$. As the biotrons responsible for foreign cells are homologous (y), even though plenty of foreign cells with heterologous genomes are available, they cannot unite and form the zygote. So they will wait for heterologous foreign cells with the X^x ovum.

Foreign cells or sperm may contain X genome or Y genome, but they are under the control of the biotron (y) from the slanted rays, produced from the male part of the planet. So the amount of energy used for expelling these foreign cells is one and the same in the bio field. So there is no possibility of releasing the X genome first. This will prevent the production of all males.

Can a foreign cell (sperm) with X genome unite with the ovum? In reality, they do not unite because X (sperm) and X (ovum) are homologous. In such a case, there is no possibility for the existence of females.

But according to our postulation, the energy form of the foreign cell with X genome that is under the control of biotron (*y*) from the male region is X^y. The ovum contains only *x* biotron which is controlled by the female region. So X genome sperm can definitely unite with X genome ova [$X^y X^x$] due to the differences in the type of biotron controlling them. As these biotrons are heterologous, XX females are produced. At this juncture, we have a question regarding the biotron concept.

XX IS ABLE TO FORM A FEMALE BUT YY IS UNABLE TO FORM A MALE

We know that in all animals, sperm with the X genome will unite with ovum of the X genome and form the zygote, which produces females. If XX = female, the YY should be male (containing two paternal genomes). Even though this sentence is logical, you cannot find a cell with two paternal genomes (YY) in the bio empire. The reason for this is given below:

Female zygote phenotype

= Female

Female zygote genotype

= XX

Energy form of biotron C type

= $X^y X^x$

Even though the genotype is homologous, the biotrons responsible for them are heterologous. So there is the possibility of a female with XX.

According to our question:

Male zygote phenotype

= Male

Male zygote genotype

= YY

Energy form of biotronic type

= Y^y

The foreign cell with two paternal YY genomes is under the control of biotrons (*yy*) from the male region. Such homologous genomes YY do not exist in the animal kingdom. Even with modern biotechnology, we cannot produce an animal with a YY genome. Under normal circumstances, males are produced with an XY genome and females are produced with XX genomes.

According to biotechnology

Male zygote phenotype

= male

Male zygote genotype

= XY

Energy form of biotron = $X^x Y^y$

In the above case, XY genomes are heterologous. The biotrons responsible for them are also heterologous. So there is a possibility of forming a male zygote with XY genomes.

The question of having a female come from the XY genome is not possible, even though the X genome is responsible for the female sex. This means all diploid XY genome cells only produce males, even though there is an X genome. There is no production of females by XY genomic cells. This can be answered by doing an experiment in physics regarding resonance.

PHYSICAL PRINCIPLES THAT SUPPRESS X-GENOME IN MALES

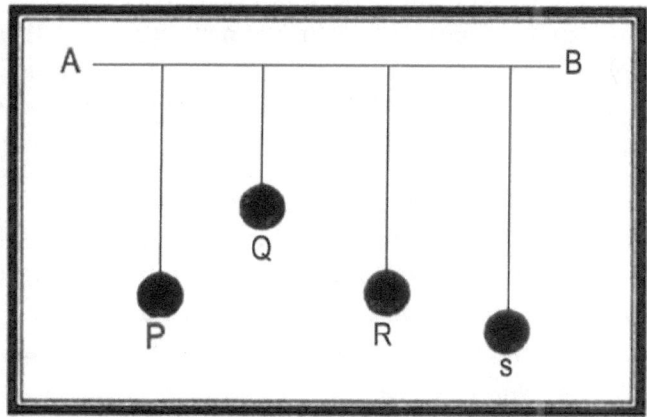

In the above diagram, four pendulums P, Q, R, and S are hung on a stretched rubber wire AB. Out of these four pendulums, two (P, R) have the same length. The other two (Q, S) have different lengths. If you draw the pendulum P to a side and let it oscillate, the rest of the pendulums will also oscillate. But Q and S pendulums will stop after some time. However, P and R will oscillate for some more time and resonate more. Now, we say P and R pendulums are under resonance.

In the above experiment, pendulums Q and S are able to vibrate. Due to the unequal lengths and unequal frequencies, they stopped resonating after some time. These vibrations are called forced vibrations.

The zygote is the source for another independent life. According to the new postulation, the zygote ($X^y X^x$) has to face a different problem, which is explained below. This is against the nature of male foreign cells.

There is no problem in the case of foreign cells formed in the female because they are all formed from the diploid genome XX. So when they are formed in females, they receive x biotrons from the female region of the planet, and the ovum contains $X^x X^x$.

But in the case of males, the foreign cells with the X genome should receive biotrons from the slant rays of the bio-waves produced from the female region of the planet. But according to our new postulate they received biotrons from the slant rays that are coming from the male region. So the waves produced in the resonance experiment are not useful. These cells are efficient enough and resonate and do all metabolic activities. Their genetic material is also reorganised according to the resonance. How far is it true? If it is completely reorganised, it would have resonated completely. Such a problem does not occur in chromosomes containing the Y genome because they are under the control of the biotrons (y) produced from the male region. So they reorganise their genetic material efficiently and do metabolic activities from the biotron (y) from slant rays with an 89°–90° angle. Let us study the situation where the zygote formed from the foreign cells formed from the $X^x Y^y$ genome.

In the case of a zygote formed from the foreign cell X^y and the ovum with an X genome, mitotic divisions occur while developing the embryo. This zygote will swallow biotrons and excite DNA, leading to splitting DNA strands and forming complementary strands. Here in this zygote, the oval part receives biotrons X^x from the female region of the planet, as in the previous case of the ovum.

If the ovum receives sperm cells with the X genome, the situation is different. Here, the X genome receives biotrons (y) from the male region of the planet, and excitation leads to the doubling of DNA.

Now, there are two options for the zygote regarding the reception of biotrons.

1. According to our postulation, the foreign cells with the X genome may receive biotrons from the male part of the planet.
2. Accepting the biotrons (x) from the female region because it has the X genome.

Out of these two possibilities, what will happen when they receive either biotron x or y?

According to the first possibility, if the zygote receives biotrons from the male region, both male and female biotrons will be present in the bio field. As these bio-waves come from different areas, they cannot accommodate and adjust in the same bio field, because they are heterologous and have opposite charges. So there is no possibility of forming bio field, stopping development in the individual. Apart from this, the biotrons that are received from the male region will have more energy due to the massive body's attraction force. So in the ovum X^x genome is dominated by the X^y and grows as male with the dominating biotrons of y in $X^x X^y$ of the zygote.

Under normal circumstances if the zygotes of XY and XX grows as male then all animal kingdom will be filled with males only. So to avoid such a situation, the sperm cells that are produced from the X genome should receive biotrons (x) from the female region.

Now, we have to know the reason for the zygote with X genome receiving biotrons from female region.

1. In an organism there is no possibility of receiving biotrons from male and female regions because they are originating from different places. Due to change in the rotational axis in the bio field, the bio field loses its coordination.
2. Even if it happens due to the presence of heterologous biotrons in a cell, their charges are nullified or neutralized, so there is no possibility of growth and development due to non-formation of the bio field.

3. In spite of having the XX genome, they will grow into male due to the biotrons (y) that are received from the male region. The biotrons from the male region are stronger than the female biotrons.

So in zygote if sperm contains X genome it should receive biotrons (X) from female region. In order to understand this problem in detail we have to study the bio field of male.

FEMALES LOSING DOMINANCE IN NATURE DUE TO GRAVITATIONAL FORCE

We know that males contain XY genomes in their cells. But we are sure that the biotrons received by it are only from the male region. However, in terms of genomes, there are two sets of genomes (XY). If the maternal genome X^x receives biotrons (X) from the female region, the paternal genome Y receives biotrons (y) from the male region. In such a case, there will be no formation of bio-field and there is no scope for further growth and development. So in the case of the male, whether it is the X genome or the Y genome, it should only receive biotrons from the male region. Only then can the Y genome resonate according to the (y) biotrons from the male region. But in the case of the sperm cell, the X genome has to be dependent on the (y) biotron. So the X genome cannot resonate according to the (y) biotrons from the male region. But the genetic material resonates to a limited extent for reorganisation. Resonance does not occur at the complete level. So in such cases, the male YY biotron dominates. Even though the X genome is present in males, it will be dormant and recessive due to the presence of Y^Y.

So nature has no partiality towards males, but instead the internal structure of the X genome is responsible for it.

DOMINANT AND RECESSIVE NATURE OF CHROMOSOMES DUE TO CHANGE IN BIOTRONS

According to the experiments of Gregor Johann Mendel, there are two factors for the determination of characteristics. These factors are called alleles. Out of these two factors, one is dominant and the other is recessive. The dominant factor is responsible for the phenotypes.

In the same way, in the case of males, XY genomes are present. Out of these two, Y dominate over the X genome, and the X genome is recessive.

But in the case of females, XX genomes are present. Both XX are resistive and not dominant. In the case of female zygotes, there are two genomes.

Energy form of the zygote that receives a biotron from the female region

$= X^x$

Energy form of the zygote that receives a biotron from the male region

$= X^y$

Out of them, X^x is from the ovum and has a homologous genome, so it can resonate freely. So, in the zygote, it will be in the dominant state. But The X^y genome received from the male is dependent on the y biotron. It has reorganised its structure to some extent with bio-waves from the male region. After entering the zygote at the time of cell division, it expects a y biotron. By that time, X^x genomes receive the ovum and can resonate with the biotron (X) from rays of the female region. There is no possibility of the existence of two different biotrons received from two different regions. So the X (sperm cell) will receive biotron X^x from female area of the planet. This will be present as a recessive female in a cell.

There is no possibility to predict if one is dominant and the other is recessive in a set of genomes. This was confirmed by Mendel's experiments.

So here, in the genome, XY, y is dominant. But the genome x acted recessive. The reason for such behaviour is explained in part V.

Q3. The majority of animals in the animal kingdom are oviparous or viviparous. In viviparous animals, some will deliver full-grown offspring, like man. But marsupials (like kangaroos) will deliver an underdeveloped foetus that will later grow in their pouches. How can this be explained by bio-waves?

Ans. In the case of any animal, the ova, sperm, or embryos are foreign in nature. This means they are not related to the same bio field. As was already explained, due to different rotational velocities, energy content, and intensity of bio-waves, there is the possibility to accumulate matter and encroach into the boundaries of the bio field by foreign cells. This is undesirable in the point of view of living beings. If something is not in accordance with the bio field, they consider them foreign and try to expel them. It can expel inert material, like phlegm, urine, or faeces, very easily by using field force. But it cannot expel either sperm or ova, which has life. By using field force instead, they can easily be expelled. In order to enhance field force, it mates with the opposite sex and enhances the bio field. As the response of the female is not as fast as that of the male, the sperm is left at the genital region of the female. Then, the embryo grows fast and enters into the periphery of the field force lines. When this embryo touches the force line of a female, the female will try to expel the foetus.

PARTURITION IN MARSUPIALS – CONTROL OF BIO FIELD

The question does not arise whether the embryo is fully grown or partially grown. In the case of a kangaroo—a marsupial species—the development of the bio field is higher at the genital region than at the head region. There is an unequal development in the bio field. The reasons for this are explained in part II chapter 14.

In the case of marsupials, the underdeveloped foetus will be expelled due to the increase in the field force at the posterior part, or the genital region.

Viviparous animals like birds and fish are under the influence of the nearest planets to the ultra-cooled massive star bodies. Due to the influence of highly energetic force lines of the massive body, outer force lines in the bio field will be strong force lines. So mass accumulates in the embryo. When this mass encroaches the external force lines of bio field, the embryo starts resisting the force by forming a strong outer shell on the egg. So there is no possibility of a zygote. This means the bio field of the egg is limited. So, with little effort of the bio field, the bird can expel its eggs very easily.

In the case of a giraffe, the development of the bio field is higher at the anterior region than at the posterior. Here also, there is an unequal development in the bio field. This is also explained in part II, chapter 14.

In the case of a giraffe, the fully developed foetus will be expelled due to the decrease in the field force at the posterior part, or the genital region.

Here, the question of the development of foetus does not arise. The moment the force lines of the bio field enter the bio field of the foetus; it will be expelled.

There is a definite reason for the development of the anterior region of the giraffe and poor development of the posterior region of the kangaroo.

This is not due to simple evolutionary process. As you are already aware, one particular animal species is under the control of one particular ultra-cooled planet or satellite body. More details are given in chapters 14, 15, and 17 in Part II.

BIOFIELDS CAUSING VARIATIONS LIKE OVIPARITY AND VIVIPARITY

There will be a number of underdeveloped, developing, and grown eggs in the genital (neutral) region of birds. All of these are not expelled at a time. Only those eggs with heavy mass that have grown enough to touch the force lines of the bio field will be expelled.

Fishes are also oviparous. Due to the influence of low energetic force lines of the massive body, which is at a distance, the force lines in the bio field will reduce. So the external force lines of fishes are stronger. Due to an increase in external field force, they will be expelled immediately in spite of less accumulation of mass in these embryos or eggs. The bio field of the egg is limited. So with the little effort of the bio field, the fish can expel its eggs very easily.

The planets that control snakes are present far behind the planets that control fishes. So they should also be oviparous.

Q4. During spermatogenesis, four sperm cells are produced from a sperm mother cell. But in the case of oogenesis, an ovum and three polar bodies are produced from the oval mother cell. Why is there such disparity? This is to be explained in terms of the concept of biotron.

BIO-WAVES CAUSING SLIGHT VARIATIONS BETWEEN SPERMATOGENESIS AND OOGENESIS

Ans. During spermatogenesis, four sperm cells are produced from one sperm mother cell.

The genotype of the sperm cells

$= Y, Y, X, X$

The biotronic type of the sperm cells

$= Y^y, Y^y, X^y, X^y$

Out of this, the *y* biotron will be energetic, as it is received from the male part that is region A of the high cooled planet or satellite body. As these four sperm cells are receiving the *y* biotron, they are more energetic and have mobility when compared to the low-energy *x* biotron. But in the case of oogenesis, during oogenesis, four ootids are produced from one oval mother cell.

The genotype of the ootids

= X, X, X, X

The biotronic type of ootid cells

= X^x, X^x, X^x, X^x

In the above case, in all four cells, the biotronic form and genotype are one and the same. All the four ootids should grow. If they are not growing accordingly, we should know the reason.

The oval mother cells are present in a null point area and do not give energy to the bio field from the received biotrons. Instead of that, these biotrons are used for absorption and duplication of DNA. In order to minimize the excitation, the excess energy will be useful for spiralising DNA.

We know that the oval mother cell contains the X_1^x, X_2^x genome. Out of this, the X_1^x genome is only from the female. It can resonate to the X biotron and show the dominant character. But this X_2^x genome, while forming the zygote (diploid state), is received from the male through sperm. In males, it is in the X^y state, and there is a reorganisation of genetic material to some extent. Later, after uniting with the female ovum, it cannot compete with X_1^x genome and receives biotron X_2^x. All these changes in the X_2^x genome are neither profitable nor lost, so the X_2^x genome shows recessive characteristics in females.

During oogenesis, the oval mother cell will receive only bio-waves of 89°–90° incline. Instead, if they receive biotrons from the BC region with 0° angles, they will be attracted towards the bio field. So chromosomes will receive bio-waves of 89°–90° angle and form chromatids with X_1, X_2, X_3, X_4 genomes. In such a case, the slant waves are very weak. In order to make biotrons from these weak bio-waves, they need additional packets of energy. So the oval mother cell should receive more energy from surrounding bio-waves and excite more by accumulating energy. This is in agreement with the formation of Henley loops and over coiling of DNA in lamp brush chromosomes of oocytes. They will receive the additional energy by exciting and resonating more. In order to excite and resonate for more energy by changing place, frequency, and intensity of bio-waves, it has to lose or exchange some part of the chromatid. So they will move towards homologous chromatids and form the bivalent genome after pairing.

Even though they appear haploid, they contain four chromatids in each bivalent genome. There will be a reorganisation of genetic material in chromatids after chiasmata formation and crossing over with homologous chromatids.

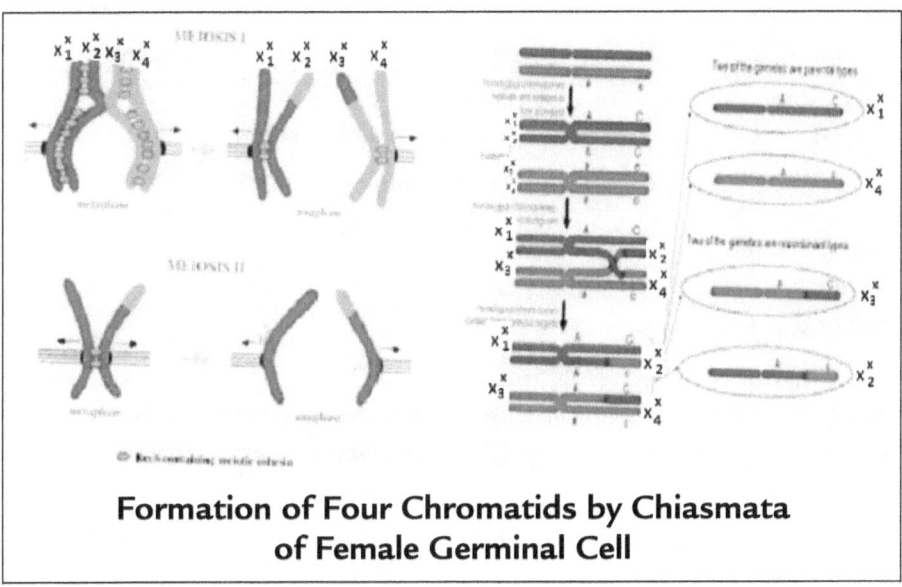

Formation of Four Chromatids by Chiasmata of Female Germinal Cell

Now, Let Us Understand Chiasmata Formation

Cell biology indicates that there is no exchange of genetic material in all four chromatids of a bivalent genome or pachytene quadrates. Only two chromatids will participate out of these four pachytene quadrates. In cell biology, this phenomenon is explained with two words: "ordinarily" and "normally."

1. Two ordinary chiasmata will be formed in a bivalent with four chromatids while reorganising genetic material.

2. Two normal chiasmata will be formed in a bivalent with four chromatids while reorganising genetic material.

There is a single meaning for ordinary or normal.

From the above picture, we know that two homologous chromosomes will pair up. There is a fair chance of forming chiasmata with X_2X_3, then X_1X_4. So in any bivalent pachytene quadrates, only two chromatids will participate in crossing over (X_2X_3), and two will be silent (X_1X_4). In such a case, the two chromatids that are not excited and resonated will form polar bodies.

Out of the two chromatids X_2X_3 that formed the chiasmata, the one derived from the mother, X_2, has to undergo reorganisation of genetic material only at certain limited places, but the other, X_3, which is derived from the male, has to undergo many modifications in its genetic material.

This is why in X_2X_3 chromatids there is less chiasmata in X_2, and X_3 needs more chiasmata for proper resonance and reorganisation of genetic material. After mating, the male and female depart. X_2 is completely reorganised with less chiasmata and resonates sufficiently, so it tries immediately to withdraw from the chiasmata at this time. There is an incomplete reorganisation of genetic material in X_3, and it is unable to resonate to the 89°–90° angle rays of the *X* biotron and convert into a polar body.

Thus, due to the non-participation of X_1X_4 chromatids, and due to incomplete participation of X_3 chromatids, they convert into polar bodies. But only X_2 reorganised completely and received a biotron from the 89°–90° angle rays, which makes it grow into an ovum.

There is a correction here. The rays that come from the female region with a slope of 89°–90° angle are very weak due to their negative attraction forces of the massive star body. Under such circumstances, the DNA will protect itself by forming through loops. All of these are possible due to the adaptations in the female bio field, and they can also receive biotrons of waves with 80°–85° angle.

In males, it takes 20 years develop and completely grow with rays of 90° angle. But in the case of females, instead of a 90° angle, at an 80°–85° angle, they will get growth and development two years earlier than males. This can be observed not only in human beings but also in other animals.

Q5. In the process of oogenesis, only chromatid X_2^x is developed into ovum out all four chromatids (X_1^x X_3^x X_4^x X_2^x). Why is it not similar in spermatogenesis? In the case of spermatogenesis, all X_1^y, X_2^y, Y_1^y, Y_2^y are developed into spermatids. The reason for such contradiction has to be answered.

Ans.

Genomic form of the four sperm cells

$= X_1 X_2 Y_1 Y_2$

Biotronic form of the sperm cells

$= X_1^y, X_2^y Y_1^y Y_2^y$

From the above equation, it can be inferred that all sperm cells receive y biotrons from the male region with more attraction forces of the massive star bodies. Unlike in the ovum, much energy is not needed for excitation of and holding biotrons in sperm cells. So with the limited energy that is available to the sperm cell $X_1^y, X_2^y, Y_1^y, Y_2^y$, they will show mobility. There is no genetic reorganisation in $X_1 Y_2$. But $X_2^y Y_1^y$ modified according to the changing situation and resonated with the rays of 89°–90° angle from the male region.

But when such sperm ($X_1^y, X_2^y, Y_1^y, Y_2^y$) reaches the ovum, only the $X_2^y Y_1^y$ sperm cell is used for zygote formation, and the rest of them ($X_1^y Y_2^y$) will perish.

Q6. The mass and volume of sperm cells is less than the ootids, even though the sperm cell's Y genome is controlled by the *y* biotron. Which is more powerful?

BIOTRONS CAUSING SIZE VARIATION – WHY SPERMS ARE SMALLER THAN OVA

Ans. The ootids contain the X genome, which will be under the control of the *X* biotron, produced from the female region of the planet with weak biowaves. Here, more energy will be spent by the genome for counteracting of the influence of the negative attraction forces of the massive star body. In order to convert the energy packets into biotron particles, it needs more energy. In order to capture such extra energy, there will be more spiralisation of DNA. If the velocity of a particle decreases, it behaves like a wave. Ultimately, more loops will be formed with single-stranded DNA, leading to more mRNA transcription. This mRNA will make more protein.

So more matter is accumulated in ova. Such a situation does not happen in spermatogenesis. There will be no loops in the chromosomes of sperm cells. The excited genomes in sperm cells lose their energy, increase the rotational velocity, and transform bio-wavic quanta into a particle. After losing excitation, the genome will be decondensed and unspiraled, and will form the chromatin network. As a mark of this, one nucleolus will be formed in the nucleus with a median notch. Through this notch, the biotron will be received from the excited genome. This is also called the nucleolar organiser chromosome. In a genome, a satellite containing chromosomes is needed for the transfer of energy.

DNA LOOPS IN LAMP BRUSH CHROMOSOMES – ROLE OF BIOTRONS AS CAUSATIVE FORCE

Both, birds and fishes are oviparous. Some birds have bigger bodies than the fishes. Out of birds, ostriches and out of all fishes, sharks are the biggest. A shark is bigger than an ostrich. Normally, bigger animals produce bigger eggs. But the ova of birds are bigger than those of fishes. What is this anomaly? This is to be explained through the concept of bio-waves.

In the case of birds, the waves that are responsible for production of ovum are (back side region of the planet or satellite) very weak, as they are under the influence of the slanting waves of 80°–90° angles. They have to take more energy in order to convert biowavic quanta into biotrons. While accumulating more energy, DNA condensation and spiralling will occur, leading to the formation of the single-stranded loops. These loops will act as a DNA template, and transcription takes place, leading to the production of more protein synthesis. That is why megalecithal eggs or large eggs are produced in the birds.

The ovum of fish is under the control of slant waves with an 89°–90° angle from the outer planets of massive body. These waves contain more energy due to less negative attraction from the outer of the massive body star. So there is no possibility for formations of loops or condensation of DNA in fishes. Thus, the protein synthesis is less than with birds, leading to the formation of microlecithal eggs, or small eggs.

The massive star body negative attraction forces are more in the case of planets, which will control birds. So to circumvent this problem, they will produce less eggs. But in the case of fishes, the massive body negative attraction forces are less due to outer orbit planets, leading to the formation of more eggs. That is why fishes produce eggs in millions, whereas birds produce eggs in dozens.

Q7. Life is considered due to the interaction of matter with bio-waves. Bio-waves are produced by the ultra-cooled celestial bodies. Astrobiology indicates that life is also an integral part of the matter or waves. There is nothing special in life except matter and its interactions with energy. Matter is always in the static or kinetic state. Inertia is considered responsible for such states. But living organisms have two states: awakened and resting. This shows that life is against the principle of inertia. This is to be explained in terms of astrobiology.

EXPLANATION ON ACTIVE AND PASSIVE STATES EXPERIENCED BY ORGANISMS

Ans. When we observe the universe, all its components, like satellites, planets, stars, galaxies, and constellations, all have rotations and revolutions. Nothing is static, and everything moves. When we observe a table in a room if it is not disturbed or moved by somebody, it will be there forever without motion. This is true. The table also moves along with the Earth at a speed of 30 km/second. Here, the observer is observing the table with the same speed and direction of the Earth, so he is not able find the change in the position of the table. But when he comes out of the reference frame of the Earth, he can find the motion of the table along with the Earth. For some reason or other, if the Earth stops its rotation for a moment, you can see the motion of a table.

Human beings have two states: awakened and resting. During day time man will be in awakened state. At night man will sleep (resting state). After taking food man will go to sleep. He cannot be in the same state for longer periods. Later he will come to awakened state and he will be in motion. All living beings only work in the awakened state to get food. Now, we have to know the reason for sleeping after eating.

You are aware that all animal beings are under the control of ultra-cooled planetary or satellite bodies. The food, whether it is vegetarian or non-vegetarian, is also formed from the ultra-cooled celestial bodies. When such food reaches the living being's gutter, it will be away from the massive body attraction forces. So he wants to sleep after eating in the awakened state. The food material that has been taken in is digested, absorbed, assimilated, and converted into energy. This energy will be useful for resisting the forces from the massive body. After some time, the energy level will be decreased. At that time, living being cannot resist the attractive forces of the massive body. So it enters the awakened state and tries to get some food.

In this way getting food and going into sleep and after some time coming to awakened state due to consumption of energy will be the cyclical activity of living being.

Normally, we don't eat more food at breakfast. We eat only a limited quantity of food for breakfast. The reason for this is simple. If you eat more for breakfast, you will be drowsy and inactive throughout the day. Without food, man cannot enter the resting state, or sleep. So in a modern society, the majority of our daily activity is linked to securing food as a daily routine.

Plants are sedentary and immobile. Plants are under the control of independent ultra-cooled nonluminous massive star bodies. The influence of planets and satellites on massive star bodies is very low (0.2%). Hence, plants are static.

Q8. Animals participate in mating to get more sexual pleasure and satisfaction. Affection, parental care, and child care are common among the animal kingdom. But our bio-wave concept says that mating occurs to expel the foreign cells from the body. Is it true?

INTERNAL FORCE LINES OF BIO FIELD GIVING ENERGY AND STABILITY TO ORGANISMS

Ans. The shape and size of an organism is dependent on the bio field. A field is nothing but the biotrons formed around the organism with force lines. These force lines are more energetic at the core and weaker at the periphery or boundary. The internal force lines are stronger than the external force lines. The external force lines of an organism are linked to massive body force lines and limbs. Due to the diffusion of external force lines, the actual energy content is less on the outside. But such things do not occur in the inner force lines.

The foreign cells (either sperm or ovum) are formed at the neutral region. The body tries to throw them out with the help of internal force lines involuntarily. Due to mating, the united field force increases, leading to strengthening the internal field force. This will work as an important factor in expelling the foreign cells. This cannot be experienced by the organism. But the impact of the external field forces can be felt very easily by an organism.

According to ancient Indian philosophies, contemplation, meditation, penance, and yoga are used for increasing the strength of the inner force lines without the influence of any external factor. Explanations for such events are not relevant here. So we are not discussing them here. The emotional factors such as love, affection, parental care, social behavior, and family system are to be explained separately.

Q9. Previously, we learnt that an organism will be under the control of biotrons, produced from a particular ultra-cooled nonluminous starbody family. It cannot receive biotrons from other areas of the same celestial body or other celestial bodies due to differential axes of biotrons, leading to the loss of coordination of the bio field. If that is true, why does a parasite live on another organism? Parasites are the integral parts of the host, such as bacteria, ascaris, tenia solium. Explain them in terms of the biotron concept.

EFFECTS OF PRESENCE OF ONE BIO FIELD (PARASITE) IN ANOTHER BIO FIELD (HOST)

Ans. All parasites cannot live in a single organism. These parasites are specific hosts/species. All parasites will have certain adaptations according to their hosts. There is a possibility of the presence of other biotrons from any ultra-cooled nonluminous celestial body, which were already present in the bio field and adjusted according to the direction and axis of biotrons that are already in the organism. Only under such circumstances can an organism live in another organism.

According to the above principle, we cannot put one organism into another organism as a parasite, even though they have the same direction and axis as the biotrons. For example, a goat cannot be inserted as a parasite in an elephant, even though they have the same direction of biotrons. This can be explained here as follows.

A magnetic substance like iron can be converted into a magnet, but not plastic. Only substances with electron domains can be converted into magnets. That is why magnetism is possible in ferromagnetic substances.

In the bio field of an organism, there are different layers. In between two layers, there are certain empty places or domains. In these domains, no biotrons are present. Only in such free empty spaces, biotrons belonging to small organisms can live without disturbing the bio field of the host. So only in a limited, specified area, one type of parasite can thrive.

There is no problem if these parasites develop their offspring by asexual methods or mitosis. But there is no possibility for development of such organisms by sexual reproduction or by meiosis. As you are already aware, the biotrons responsible for gametes or haploid cells are against the bio field of the mother with a 90° angle.

So these parasites have to select another organism or host, which can accept biotrons of the haploid gametes for sexual reproduction. This means there is a need for another host for sexual reproduction. This can be seen in a majority of parasites.

In the case of malarial parasites, asexual reproduction will be completed in man. But sexual reproduction takes place in female anopheles mosquitoes. The situation in humans does not allow for the growth of gametes, because the gametes are exposed to bio-waves produced from the 90° angle. That is why they migrate to female anopheles mosquitoes. These parasites are under the control of biotrons produced from the rotating small planetary body. Due to a change in its direction after some time, the field in the female anopheles mosquito will not be congenial for their growth. So they migrate to man for further development.

The surface area of the ultra-cooled nonluminous celestial bodies can be divided into latitudes and longitudes. The bio-waves that are produced from a particular latitude will control one particular cluster of people on the Earth. All human beings are under the control of one particular ultra-cooled planetary body. Yet there are differences in shape, size, structure, and races, and subtler variations in characteristics of human beings. For example, Caucasoids, Negroids, Australoids, Mongoloids, etc. The variations in the latitudes and longitudes the ultra-cooled planetary body is responsible for variations among the human beings. In the same way, the parasites residing in a particular race are different from the other race. For example, tropical parasites are different from temperate parasites. Certain parasites are endemic to certain ethnic races.

Q10. Biology tells about the phenotype and genotype of a species. But astrobiology added one more form, called the biotronic form. Can we understand the complete nature of an organism with these three forms, or do we need another form?

TOTAL REALISTIC STATE OF ORGANISM EXPLAINED BY FOUR TYPES

Ans. The genotype and phenotype tells about the external and internal physical status of the organism. But the biotronic form tells about the energy status of an organism. This biotronic form also cannot completely describe an individual because that is not static. Due to changes in the planetary rotation, there is growth, development, and senescence in the individual. For example, a 12-years age boy, a 25-years age young man, and a 90-years age old man belong to the same phenotype, genotype, and biotronic form. But we know that not all are in the same state. So one more form is needed to explain the nature of an organism. That is called the biotronic angular form. This biotronic angular form tells about the area of the planet from where it receives biotrons, and with what angle it is receiving the bio-waves.

There are three states in the biotronic angular form. They are:

1. Growth phase
2. $= +$
3. Static phase
4. $=$ no symbol is needed
5. Reduction phase
6. $= -$

Biotronic angular form of sperm cell

$$= X^{y+90°} \text{ or } Y^{y+90°}$$

Biotronic angular form of ovum

$$= X^{x+81°}$$

Let us take man as an example. The different forms are given below:

Human Beings

MALE				
Age	Phenotype	Genotype	Biotronic Form	Biotronic Angular Form
DAY 1	MALE	XY	$X^y Y^y$	$X^y Y^y$ +86.15°
YEAR 1	MALE	XY	$X^y Y^y$	$X^y Y^y$ +81.65°
10 YEARS	MALE	XY	$X^y Y^y$	$X^y Y^y$ +41.5°
20 YEARS	MALE	XY	$X^y Y^y$	$X^y Y^y$ +0°
30 YEARS	MALE	XY	$X^y Y^y$	$X^y Y^y$ +0°
40 YEARS	MALE	XY	$X^y Y^y$	$X^y Y^y$ +0°
60 YEARS	MALE	XY	$X^y Y^y$	$X^y Y^y$ +0°
82 YEARS	MALE	XY	$X^y Y^y$	$X^y Y^y$ −9°
90 YEARS	MALE	XY	$X^y Y^y$	$X^y Y^y$ −45°
FEMALE				
Age	Phenotype	Genotype	Biotronic Form	Biotronic Angular Form
DAY 1	FEMALE	XX	$X^x X^x$	$X^x X^x$ +77.15°
YEAR 1	FEMALE	XX	$X^x X^x$	$X^x X^x$ +72.65°
10 YEARS	FEMALE	XX	$X^x X^x$	$X^x X^x$ +32.5°
20 YEARS	FEMALE	XX	$X^x X^x$	$X^x X^x$ +0°
30 YEARS	FEMALE	XX	$X^x X^x$	$X^x X^x$ +0°
40 YEARS	FEMALE	XX	$X^x X^x$	$X^x X^x$ +0°
60 YEARS	FEMALE	XX	$X^x X^x$	$X^x X^x$ +0°
82 YEARS	FEMALE	XX	$X^x X^x$	$X^x X^x$ −18°
90 YEARS	FEMALE	XX	$X^x X^x$	$X^x X^x$ −54°

Females receive energy packets from 81° slanting rays instead of 90° rays due to the convenience of the bio field and establish biotrons with the additional amount of energy produced from the resonance of chromosomes, leading to early development. That is why females will develop and mature two years earlier than males. For complete development in males, it takes 20 years, whereas in the case of females it is only 18 years. This early development of females is not only in human beings but also in other animals.

Ova of women develop with biotrons from 80°–90° rays.

Ova of birds develop with biotrons from 80° rays.

Ova of tetrapods develop with biotrons from 81°–85° rays.

Ova of fishes and amphibians develop with biotrons from 85°–88° rays.

Q11. In the animal kingdom, all somatic cells are diploid. If it is female, it contains XX and male XY. According to the bio-waves theory, one genome is linked one biotron. If there are two biotrons in a cell, both should have the same place, phase, and direction. Accordingly, males had $X^y Y^y$ and females $X^x X^x$. If there is no possibility for the union of homologues biotrons at the time of mating, how will diploid cells have two homologous biotrons ($X^y Y^y$ or $X^x X^x$) as a basis for two genomes? This has to be explained.

Ans. Previously, we proposed the concept of one genome, one biotron, and one nucleolus. If it is true, how does Acetabularia have hundreds of nucleoli in a cell? In all organisms, during old age, there will be a reduction or disappearance of the nucleolus. It means there are some cells without the nucleolus and biotrons. Is it possible? This has to be clarified.

Biotronic Form of Different Gametes

In human beings, three types of gametes will be formed. They are:

1. X ovum
2. X sperm
3. Y sperm

There is a possibility of union of these three gametes in factorial 3 ways. Factorial $3 = 3 \times 2 \times 1 = 6$. The permutations and combinations are:

S. No.	Gamete of First Genome	Gamete of Second Genome	Status of Genomes Participating in Union	Status of Sex	Resultant Zygote
1.	X sperm	X ovum	Homozygous	Hetero sex	XX female
2.	Y sperm	X ovum	Heterozygous	Hetero sex	XY male
3.	X sperm	X sperm	Heterozygous	Homo sex	No zygote
4.	X sperm	Y sperm	Homozygous	Homo sex	No zygote
5.	Y sperm	Y sperm	Homozygous	Homo sex	No zygote
6.	X ovum	X ovum	Homozygous	Homo sex	No zygote

In the first (homozygous) and second instances (heterozygous), only zygotes are formed. In both cases, the sexual status is heterosexual. In the other four cases, even though the genomes are heterozygous, the status of the sex is homosexual. That is why a zygote is not formed with them.

In reality, the X sperm and Y sperm are heterologous genomes. The sperm with an X genome contains the maternal genome. The sperm with a Y genome contains paternal genome. The formation of zygotes is possible only with the union of male and female gametes. But there is the possibility for the formation of a zygote with the X ovum (maternal genome), with the X sperm (maternal genome), and the production of the female. How is it possible for the union of maternal genome (X) with maternal genome (X) and the production of a zygote XX (female)?

BIOTRONIC TYPE REGULATING THE FATE OF DIFFERENT GAMETES

This can be explained on the basis of biotrons. There are two types of biotrons:

1. Y biotrons are released from the male part Y region of the planet.

2. X biotrons are released from the female part X region of the planet.

Gametes:

X sperm cell, Y sperm cell, X ovum

Biotronic forms:

$X^y \, Y^y \, X^x$

S. No.	Gamete of First Genome	Gamete of Second Genome	Biotronic Form	Status of Biotron Sex Participating in Union	Genome of Sex Status	Status of Organism Sex Produced	Resultant Zygote
1.	X sperm	X ovum	$X^y \, X^x$	Heterozygous	Homo sex	Hetero sex	XX female
2.	Y sperm	X ovum	$Y^y \, X^x$	Heterozygous	Hetero sex	Hetero sex	XY male
3.	X sperm	Y sperm	$X^y \, Y^y$	Homozygous	Hetero sex	Homo sex	No zygote

S. No.	Gamete of First Genome	Gamete of Second Genome	Biotronic Form	Status of Biotron Sex Participating in Union	Genome of Sex Status	Status of Organism Sex Produced	Resultant Zygote
4.	X sperm	X sperm	$X^y X^y$	Homozygous	Homo sex	Homo sex	No zygote
5.	Y sperm	Y sperm	$Y^y Y^y$	Homozygous	Homo sex	Homo sex	No zygote
6.	X ovum	X ovum	$X^x X^x$	Homozygous	Homo sex	Homo sex	No zygote

From the above table, we come to the conclusion that the biotrons participating in the union should be heterologous. There is no incompatibility for homologous genomes.

There is no possibility for the formation of a zygote from the union of two homologous biotrons. We have to explore the reason for such a situation in terms of physics.

Ans. Biotrons are formed from protomagnetic waves. According to quantum physics, these protomagnetic waves travel in the form of energy packets. That is why we have coined the term biotrons for these energy packets. According to the wave property, the biotrons also revolve. This revolution may be in the clockwise or anticlockwise direction. As these biotrons are produced from the protons, they may have charge and field. Any revolving particle will have dipoles. In reality, all biotrons will have the same structure. They are not different. We have to know what homologous and heterologous biotrons are.

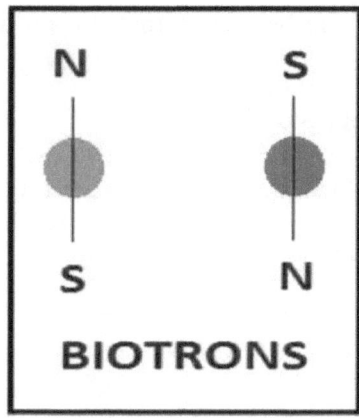

This can be explained by taking two biotrons as an example.

In the above picture, two biotrons revolve anticlockwise. Due to this revolution, dipoles are formed. They are called north (N) and south (S) poles. As these biotrons revolve in the same direction,

$$\begin{matrix} N & N \\ \cdot & \cdot \\ \cdot & \cdot \\ S & S \end{matrix}$$

They are homologous and repel each other. So there is no possibility of these two biotrons getting closer. So they revolve separately due to repulsion forces.

In the above case, two biotrons are present. A biotron revolves anticlockwise, and the B biotron revolves clockwise. Due to revolution, dipoles are formed. As these biotrons are not revolving in the same direction, they are heterologous $\genfrac{}{}{0pt}{}{n\ \ S}{\underset{S\ \ N}{1\ 1}}$ and they attract each other. So there is the possibility of these biotrons getting closer. This will lead to the increase of linear velocity. Due to friction, revolutionary velocity decreases to the biotrons. If revolutionary velocity decreases, it changes from a particle to a wave, so they leave the cell.

From this, we can learn that, due to the nearness of homologous biotrons, the revolutionary velocity does not change. But when two heterologous biotrons get close, the revolutionary velocity decreases and they change from particle to wave.

Now, let us discuss the bio-waves that are produced from a planet in terms of the above concept.

The Y biotrons produced from the male region with a 90° angle are heterologous to the x biotrons produced from the female region, with a 90° angle due to a change in the direction, place, and polarity.

Let us consider gametes and observe the formation of zygotes:

Biotronic form of X sperm $\genfrac{}{}{0pt}{}{n}{\underset{S}{1}}$

$= X^Y$ {it is correct}

Biotronic form of Y sperm $\underset{S}{\overset{n}{1}}$

$= Y^Y$ {it is correct}

Among them, $Y\ Y$ biotrons are homologous. Due to repulsion of $\underset{S\ S}{\overset{N\ N}{:\ :}}$ there is no possibility of forming a zygote. Now, let us consider X ovum and Y sperm.

There biotronic forms are $\underset{N}{\overset{S}{1}} X^X\ \underset{S}{\overset{n}{1}} X^Y$

As they are heterologous biotrons due to attraction of $\underset{S\ N}{\overset{n\ S}{1\ 1}}$ they will get closer to each other and a zygote will be formed.

Due to the decrease in revolutionary velocity of biotrons (due to friction) in a heterologous pair of biotrons, it changes its nature from a particle to a wave and escapes from there. Such a situation is possible in zygotes. To counteract this situation, there are three possible ways to the genomes:

1. **Leaving the Biotron from the Zygote.** If the cell loses biotrons, it will lose energy. This is suicidal.

2. **Capturing the Biotron.** There is no energy provided to the genome to prevent the attraction forces between two biotrons. The matter present in the genome cannot capture the energy in the form of a wave. If possible, it has to capture energy packets of the wave form and convert them into matter.

3. **Absorbing the Biotron.** The genome always tries to capture a biotron that is escaping to use for increasing the internal energy of DNA, leading to the splitting of the DNA strand and the formation of a complementary DNA strand.

From the above analysis, it is imminent for cell division in the zygote. In the zygote, two heterologous biotrons and two genomes are present. Due to DNA auto-catalysis, two genomes will become four genomes. Out of these four, two are X genomes and have $\underset{N}{\overset{S}{1}} X$ and another two are Y genomes $\underset{N}{\overset{S}{1}} Y$.

In total, there are four genomes (X, X, Y, Y), and four biotrons (X, X, Y, Y) are present in the zygote. Out of these four biotrons, X, y are attracted and divided into two cells ($X^x\ Y^y$, $X^x\ Y^y$). Due to the presence of heterologous biotrons in each cell, they will go on dividing cyclically. Thus, the number of cells increases in the embryo and matter increases. This embryo is growing at the neutral area of the null point. After some growth of the foetus, the foetus touches the force lines of the bio field. In response to this, the force lines of the bio field start resisting the entry of the foetus into its periphery. At that time, one important change occurs in the foetus cells.

To explain the above phenomenon, let us take the example of a zygote with the XY genome. After the embryonic growth, all cells in the foetus contain X^xY^y biotrons. In order to resist the force of the bio field of females, the embryo has to increase the energy content of its biotrons. The X biotron has less energy than the y biotron.

In order to increase the field force of the embryo, the X^x genome will convert into X^y. Here, after the biotronic form of all cells in embryo change from X^xY^y to X^yY^y, the callus-like embryo will transform according to the shape of the bio field. It becomes male, as it has a y biotron with a Y genome. Here after the X genome will be recessive and Y genome will be dominant. After accepting y biotron by the X genome.

Now let us study the development in the zygote with XX genome. The biotronic form of this cell is X^xX^y. The X biotron of the X genome came from ovum will resonate efficiently because X genome is reorganised according to the x biotron. In order to increase the energy it has to accept x biotron of XY. But the genome is not modified according to Y biotron. So in order to increase the resonance, it will only go for x biotrons. In the same way, the X genome from the ovum also reorganises according to the x biotron. Even though more energy is present in the y biotron, it cannot receive it because the DNA is not formed from the y biotron but organized according to the x biotron.

After that, the cells in the callus (X^xX^y) will transform into X^xX^x. Now, all the biotrons of the embryo are of the same type. So with the united force of all the cells of the embryo, it will resist the female bio field force exerted on the force lines at the periphery of the null point or neutral area. More details are given in part-II. After studying this knowledge will arise a basic question, that is Human species controlling planetary body is one of the high cooled planet and satellite bodies which is constructing the Animelia species so there is no speciality to the human race planet, but on earth Human species is highly intelligent and acquired so many knowledge, so how is it possible?

REASONS FOR HIGHLY INTELLIGENCE OF HUMAN SPECIES AMONG THE ANIMAL SPECIES

Ans. To clarify this question we have to understand the refraction nature of the waves. Refraction is the change in direction of the wave propagation due to a change in its transmission medium. For example, when we travel in a hot road, we can see a sheet of water at a distance but children think there is a lot of water. We know that it is not true.

When we place a small coin in a glass with water and observe the place of it in glass and then when we try to touch the coin where it appears to be, we will fail and realise there is no coin, where it is appearing. These two incidents occur due to the wave character of refraction. In above incidents we are inducted by what we perceive is true but when respond to that we fail. This means induction is not correct and the response failed so there is no knowledge gained by preceptors.

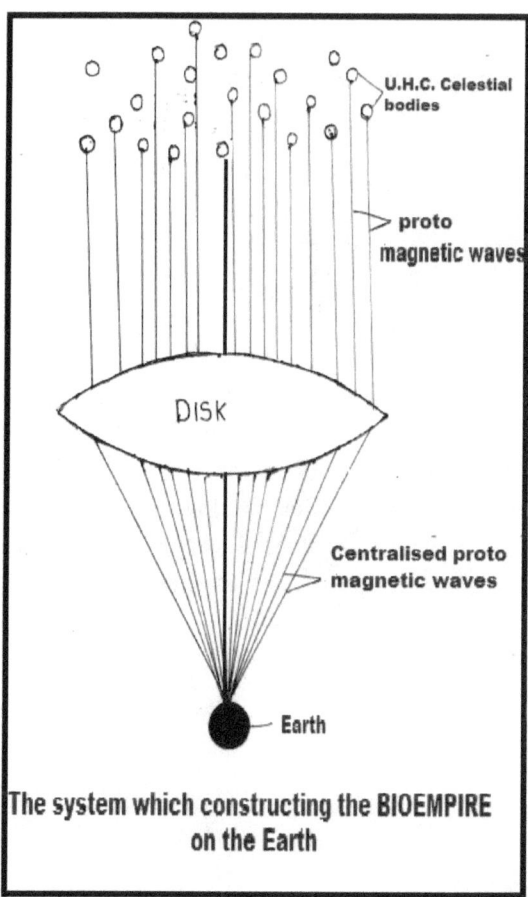

The system which constructing the BIOEMPIRE on the Earth

In the above diagram all the bio waves coming from high cooled celestial bodies entering in the disk are effecting by disk and changes their direction and centralising on earth except one celestial body that is in centre of the disk, it escapes the character of the refraction and appearing its direction is not changed. Now that celestial body belongs to Human species. The total Animal species living beings except human species are receiving false inductions on earth and they are failing in response. Hence they are not gaining any knowledge and not showing intelligence like human beings. When induction is correct response also is correct and success remains and then a bit of knowledge acquired.

Another thing is all the high cooled celestial bodies, disk and earth are not static. They are all moving with a great speed so our human species celestial body also exits some time from the centre point of the disk and may occupy other near celestial body. Then the bio waves of the human species celestial body also effects by refraction. Then all human beings perceive their responses are not correct according to their inductions so human race knowledge is disturbed and behaves like other animal species living beings. This means all civilizations constructed by human species intelligence disappeared.

Conclusion: Taking a biotron concept, I gave full explanation about bio-empire at all levels i.e, from Kingdoms-Genus and species-living being-organs-tissues-cell-genome-gene and quanta levels and also explained the characteristics of life forms namely; birth-growth-death-feeding-reproduction-sex diversifications-sizes of body and so many variations through this biotron concept. If we think this concept may be wrong, it is impossible to travel in bio-empire—such a long journey with this concept without any deviation. I am also humbly saying that it was possible for me after spending more than 25 years in my life on this concept. I hope the other parts—Astrobotany, Astro micro biology and Modern astrobiology will reach the readers at an early date. All literates, have to necessarily acquire this knowledge. so, I request the reader to share this knowledge with others friends and other persons in the society.

www.ingramcontent.com/pod-product-compliance
Lightning Source LLC
Chambersburg PA
CBHW030749180526
45163CB00003B/952